Foreword

The organic chemical industry has grown tremendously since World War II. The manufacture of inorganic chemicals accounted for only 15 per cent of the total turnover of the industry in 1968 while organic chemicals accounted for about one-third of the turnover; this is the reverse of the position existing at the beginning of this century. Industrial organic chemistry should, therefore, be given at least as much coverage in A-level textbooks as is given to industrial inorganic chemistry.

One of the main objects of the first edition of the two monographs on industrial chemistry was to bring certain sections of industrial chemistry dealt with in O and A-level textbooks of chemistry up to date. This object has been kept in mind in preparing this revised edition. However, a much wider approach has been adopted which, it is hoped, will extend the range of usefulness of the monograph to post A-level courses such as those leading to a Higher National Certificate and Diploma in Chemistry and Applied Chemistry.

A few sections of the first edition have been omitted and others expanded, *e.g.* the production of aromatic compounds from coal and petroleum, and the chapter on carboxylic acids. Several new sections have been introduced including an introductory chapter on raw materials for the production of heavy organic chemicals and one on high polymers and surface-active agents. It was thought advisable to include the latter chapter in view of the frequent references which have to be made to plastics, synthetic fibres, synthetic rubbers and surface-active agents. However, it is stressed that the treatment of high polymers is essentially an introductory one. A number of flow diagrams have been included, uses of important compounds have been compiled in tabular form and statistics of their production given wherever possible. The current prices of heavy organic chemicals and a map showing the distribution of chemical works and oil refineries in the UK are included in the Appendix.

I am very grateful to the following for providing me with literature describing the manufacture and/or uses of chemicals: BP Chemicals International Ltd; Badger Co. Ltd; Chemical Construction Corporation; Esso Petroleum Co. Ltd; Gulf Publishing Company, Houston, Texas; Humphreys and Glasgow Ltd; Kellogg International Corporation; and Simon–Carves Chemical Engineering Ltd.

D. M. Samuel
Gloucester,
September 1971

Contents

Monographs for Teachers

This is another publication in the series of Monographs for Teachers which was launched in 1959 by the Royal Institute of Chemistry. The initial aim of the series was to present concise and authoritative accounts of selected, well-defined topics in chemistry for the guidance of those who teach the subject at GCE Advanced level and above. This scope has now been widened to cover accounts of newer areas of chemistry or of interdisciplinary fields that make use of chemistry. Though intended primarily for teachers of chemistry, the monographs have proved of great value to a much wider readership, including students in further and higher education.

13/661.8

PART 3: AROMATIC COMPOUNDS

1. Raw Materials for the Production of Heavy Organic Chemicals

Introduction

Although much of this monograph deals with the manufacture and uses of heavy organic chemicals, it should be remembered that the organic chemical industry includes industries manufacturing a wide variety of other substances such as plastics, synthetic fibres, synthetic rubbers, surface-active agents, surface-coatings, dyestuffs and organic pigments, explosives, pharmaceuticals, pesticides and other agricultural chemicals, glue and gelatine, foodstuff additives, refrigerants, photographic chemicals, perfumery ingredients and synthetic flavours. Most of these substances are referred to in the monograph, but in view of the quantities of heavy organic chemicals involved in their manufacture, it has been considered advisable to include an outline treatment of plastics, synthetic fibres, synthetic rubbers and surface-active agents. Thus, world production of plastics materials (the plastics industry is little more than a century old) is expected to increase from the present 30 Mt to 90 Mt by 1980. The industries manufacturing these products employ large numbers of chemically trained personnel. Plastics, petroleum-derived organic chemicals and pharmaceuticals represent the fastest growing sectors of the chemical industry. About 15 per cent of all the employees in the chemical industry are involved in the manufacture of pharmaceuticals.

The basic raw materials from which heavy organic chemicals are produced are petroleum and natural gas, coal, agricultural products and animal and vegetable fats. Coal was by far the most important raw material in the UK before World War II. Petroleum products were first used as feedstocks during the war years and in 1968 in the UK they accounted for about 85 per cent (by weight) of the total production of organic chemicals[1] and 90 per cent of the production in the US. It has been forecast that by 1975 the production of petroleum-derived organic chemicals in Western Europe will amount to 30 Mt/a (production amounted to only 100 000 t/a in 1950). Aromatic hydrocarbons were formerly made almost exclusively from coal, but by the late 1970s it is expected that about 80 per cent of the requirements of the aromatic hydrocarbons benzene, toluene and xylenes will be derived from petroleum. The use of coal as a

1

raw material for the production of organic chemicals is discussed on p 94 and of the other raw materials at the end of this section. World trade in primary hydrocarbons, intermediates, and polymers in 1970 was worth nearly £3000 m.

Petroleum[2]

World production of crude petroleum exceeded 2000 Mt in 1968. In 1969, consumption of petroleum in the UK amounted to more than 89 Mt. Sufficient North Sea oil should become available in the next few years to satisfy about 50 per cent of the UK's petroleum requirements. Crude oils consist essentially of a complex mixture of hydrocarbons which may be predominantly paraffinic or naphthenic (cycloparaffinic) in nature, or they may contain substantial quantities of both types of hydrocarbons. The processing of crude petroleum begins with fractional distillation at or above atmospheric pressure in continuous pipe stills. In a pipe still the oil is pumped through banks of piping heated by radiation from burning fuel oil, which brings the temperature of the feedstock up to about 400 °C. About three-quarters of the oil vaporizes. The mixture of liquid and vapour then passes into a fractionating column. *Figure 1* illustrates primary distillation. The crude oil is split up into the following fractions (given in order of decreasing volatility):

1. Methane, ethane, propane and butanes
2. Light gasoline
3. Heavy gasoline or naphtha
4. Kerosine (maximum boiling point 275–280 °C)
5. Light gas oil
6. Residual oil

Further fractional distillation of residual oil (some of which may be incorporated into fuel oils) is carried out under reduced pressure to yield heavy gas oils, fractions suitable for the production of gasoline by fluid catalytic cracking or for the production of lubricating oils, and a residue which may consist of bitumen (depending on the type of crude oil being distilled) or may only be suitable for use as fuel oil. In Europe more than 90 per cent of the production of crude oil is used as fuel and only about 3 per cent as feedstock for the production of chemicals. Nevertheless, UK sales of petroleum-derived organic chemicals in 1969 amounted to 5.3 Mt.

The naphtha fraction, which has too low an octane number to allow its incorporation into motor spirit, is used mainly as feedstock for catalytic reforming units (p 98) to produce high-octane blending components (reformate) for motor spirit and for producing town gas

---

Fig. 1.   (Facing page) Primary distillation. (By courtesy of Esso Petroleum Co. Ltd.)

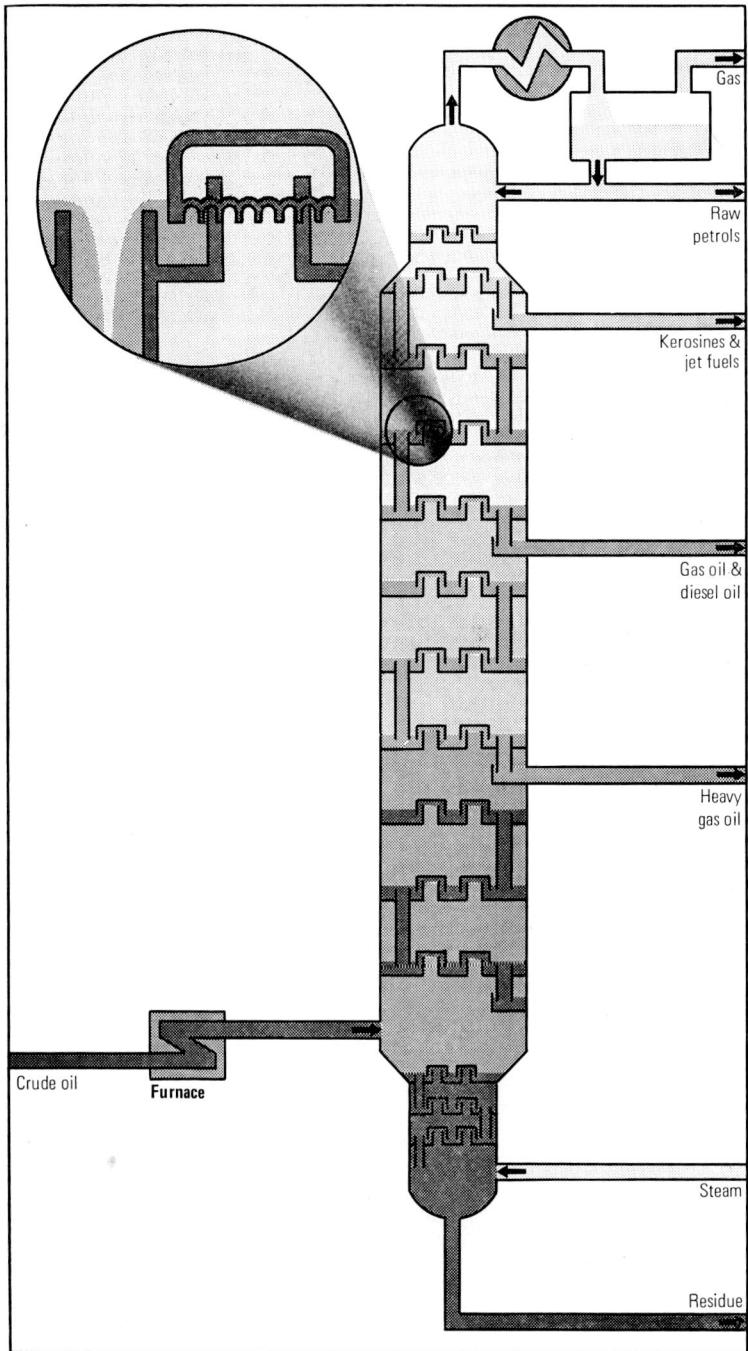

Gas

Raw
petrols

Kerosines &
jet fuels

Gas oil &
diesel oil

Heavy
gas oil

Crude oil   **Furnace**

Steam

Residue

by pressure steam-reforming.   Large quantities of reformate are also used for the production of pure aromatic hydrocarbons (p 98).   Naphtha is also used in large quantities as a feedstock for the production of olefins by steam-cracking (see later); because of their reactivity the latter are by far the most important raw materials for the production of petroleum-derived chemicals.   In recent years naphtha has also been used in the UK as a feedstock for the production of acetylene.   UK consumption of naphtha in 1969 exceeded 11 Mt (an increase of 1 Mt on the 1968 figure), about half of which was used as feedstock for the production of organic chemicals.

Olefins are also produced in the course of fluid catalytic cracking processes (so called because the oil vapour is in contact with finely divided catalyst, and the mixture behaves like a fluid) applied to feedstocks such as vacuum distillates from residual oil to produce motor spirit blending components.   The most common reactions which take place in the course of catalytic cracking processes are fission of the carbon chain to give olefins and paraffins, *e.g.* $CH_3(CH_2)_{10}CH_3 \rightarrow CH_3CH\!=\!CH_2 + CH_3(CH_2)_7CH_3$, and dehydrogenation, $C_nH_{2n+2} \rightarrow C_nH_{2n} + H_2$.   Some dehydrocyclization of paraffins also takes place (p 101).   Long-chain (C_6–C_{18}) α-olefins (*i.e.* those having the double bond at the end of the carbon chain) have long been produced in the UK by the thermal cracking of slack wax (a mixture of oil and waxes of low wax content which separates when lubricating oil fractions are diluted with a solvent and chilled).

FIG. 2.   Distillation of crude oil.   (By courtesy of Esso Petroleum Co. Ltd.)

The process is carried out in the vapour phase in the presence of steam. Long-chain α-olefins are used for the production of sodium secondary alkyl sulphates and higher alcohols. Consumption of petroleum-derived feedstock (other than naphtha) for the production of chemicals in the UK in 1969 amounted to 390 000 t.

The lower paraffins derived from petroleum have been little used in the UK as sources of heavy organic chemicals. Propane and the butanes in the form of LPG (liquefied petroleum gases) are used almost entirely as fuels, propane also being used for metal-cutting purposes. North Sea natural gas, consisting almost entirely of methane, will be used largely as a fuel but also as a feedstock for the production of synthesis gas (*i.e.* a mixture of carbon monoxide and hydrogen) to make ammonia, methanol and higher alcohols (by the OXO or hydroformylation process). It can also be used instead of naphtha for the production of acetylene. Other uses of the lower paraffins are discussed later. Long-chain n-paraffins are separated from accompanying branched-chain paraffins and cycloparaffins by two processes (*see* p 37). Slack wax, referred to in the previous paragraph, is a valuable feedstock for the production of n-paraffins which are used for making biodegradable detergents (p 32) and secondary plasticizers for polyvinyl chloride (p 20).

Most petroleum-derived heavy organic chemicals are produced in continuous, large scale, automatically controlled plants. Production on the largest scale is essential because of the high capital cost of the very complex petroleum-derived chemical plants. Thus the capacity of an olefins plant operated by ICI is 450 000 t/a of ethylene. It will be apparent that processes which employ the least number of stages will be the most economical. Thus, until comparatively recently acetic acid was made from ethyl alcohol (produced from ethylene derived from naphtha) *via* acetaldehyde:

$$C_2H_4 \xrightarrow{\text{H}_2\text{O}} CH_3CH_2OH \xrightarrow{-2\text{H}} CH_3CHO \xrightarrow{\text{O}_2} CH_3COOH$$

It is now made by liquid-phase oxidation of paraffins derived from naphtha (p 77). Production in large plants, coupled with the low cost of the raw materials, has stimulated the use of petroleum-derived chemicals for the manufacture of plastics, synthetic rubbers, synthetic fibres, surface-active agents, surface-coating ingredients, paper coatings, adhesives, fertilizers and other agricultural chemicals.[2] The demand for these materials is ever-increasing since they (apart from the agricultural chemicals) are used in the production of consumer products such as motor cars (the average motor car contains about 30–40 lb of plastics). Nearly all modern plastics are, in fact, based on petroleum-derived chemicals. A number of examples are given later which illustrate the extent to which petroleum-derived chemicals are fulfilling a demand which can only partially be met by the use of other raw materials (*e.g.* production of coal-tar chemicals is limited by the demand for coke, which in turn depends on steel production). A number of chemicals are more cheaply produced from petroleum than from other raw materials, *e.g.* cyclohexane, required for the production of nylons 6 and 66, is more cheaply produced from petroleum-derived benzene than from benzene derived from coal by carbonization. Amongst other obvious advantages of petroleum over coal may be mentioned the greater price stability of petroleum products and the greater ease of handling (by pumping).

Production of olefins

Lower olefins (ethylene, propylene, C_4- and C_5-olefins) are produced in oil refineries in the course of fluid catalytic cracking operations to produce blending components for motor spirit. However, the demand for the lower olefins, particularly ethylene and propylene, has grown so large that cracking processes have been developed to maximize yields of these two olefins, and the most widely used process in Europe is the steam cracking of petroleum naphtha. Naphtha cracking now accounts for about 80 per cent of the ethylene produced in Europe.[3] In the US, however, where ample supplies of 'wet' natural gas (*i.e.* natural gas containing appreciable quantities of paraffins up to C_7) are available, ethane and propane are the chief sources of ethylene and propylene.

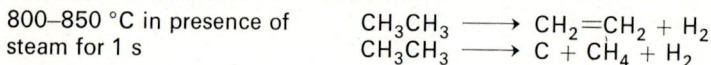

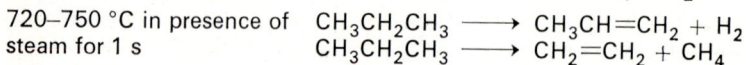

800–850 °C in presence of steam for 1 s

$$CH_3CH_3 \longrightarrow CH_2{=}CH_2 + H_2$$
$$CH_3CH_3 \longrightarrow C + CH_4 + H_2$$

720–750 °C in presence of steam for 1 s

$$CH_3CH_2CH_3 \longrightarrow CH_3CH{=}CH_2 + H_2$$
$$CH_3CH_2CH_3 \longrightarrow CH_2{=}CH_2 + CH_4$$

In the cracking of propane the propylene output is about 40 per cent by weight of the yield of ethylene.

Petroleum naphtha is obtained by fractional distillation of crude oil at atmospheric pressure, and has a boiling range 50–200 °C. It consists largely of paraffins with up to about 20 per cent of naphthenes (cycloparaffins) and a smaller percentage of aromatic hydrocarbons. Since ethylene is in greater demand than propylene, cracking is carried out under conditions which will give more than twice as much ethylene as propylene; yields in excess of 30 per cent ethylene are obtained. Under more severe cracking conditions, *i.e.* at higher temperatures and for longer cracking times, the yield of ethylene is increased at the expense of propylene and C_4-olefins, and considerably more methane (15–20 per cent) and hydrogen result. The most important constituent of the C_4-fraction is butadiene (3–5 per cent) which is in great demand. In a typical steam-cracking process the ratio of steam to naphtha is 0.3–0.6 lb: 1 lb. The steam reduces the partial pressure of the naphtha vapour so that the cracking is, in effect, carried out under reduced pressure, thus increasing the yield of useful products. It also reacts with any carbon formed:

$$C + H_2O \longrightarrow CO + H_2$$

The mixture of naphtha and steam is pumped rapidly through the tubes of the cracking furnace which are heated by radiation from burning fuel oil. Typically the mixture may be at 810 °C for about one second. The effluent gas from the cracking furnace is cooled rapidly and the heat recovered is used to generate high-pressure steam. The gas is then passed to a fractionating column; from the base of this a liquid fraction is withdrawn which, on further fractionation, yields an aromatics-rich product (some aromatic hydrocarbons were present originally but they are also formed by the dehydrogenation of cycloparaffins and by the cyclization of paraffins followed by dehydrogenation) suitable for blending into motor spirit. The gaseous effluent from the head of the column is compressed, washed with caustic soda solution to remove carbon dioxide and hydrogen sulphide, dried and cooled. The liquefied portion is then partially separated into its constituents in a series of fractionating columns operated under pressure. Each column is provided with a reflux condenser through which a refrigerant is circulated; for example, liquid ethylene is used as coolant in the de-methanizer column, *i.e.* the column from which methane is removed. The products obtained are 95 per cent methane, C_2-hydrocarbons, 90–94 per cent propylene and a C_4-fraction consisting of butanes, butenes and butadiene. The C_2-fraction is treated with hydrogen in the presence of a catalyst under conditions which convert acetylene to ethane and is then fractionated to give ethylene and ethane. The C_4-fraction is separated into isobutene, isobutane, butadiene and n-butenes (p 40). A flow diagram for a modern ethylene plant is shown in *Fig. 3.*

Other raw materials

Agricultural products used for the production of organic chemicals include wood (the source of cellulose), potatoes and cereal crops (sources of starch), sugar cane and sugar beet, corn cobs, animal and vegetable fats, hides, bones and milk.

Wood consists of cellulose, hemicelluloses and lignin. Chemically-produced wood pulp is obtained by dissolving out the non-cellulosic constituents by means of calcium bisulphite solution containing sulphur dioxide, dilute sodium hydroxide solution or a solution of sodium hydroxide and sodium sulphide. Treated pulp is used in

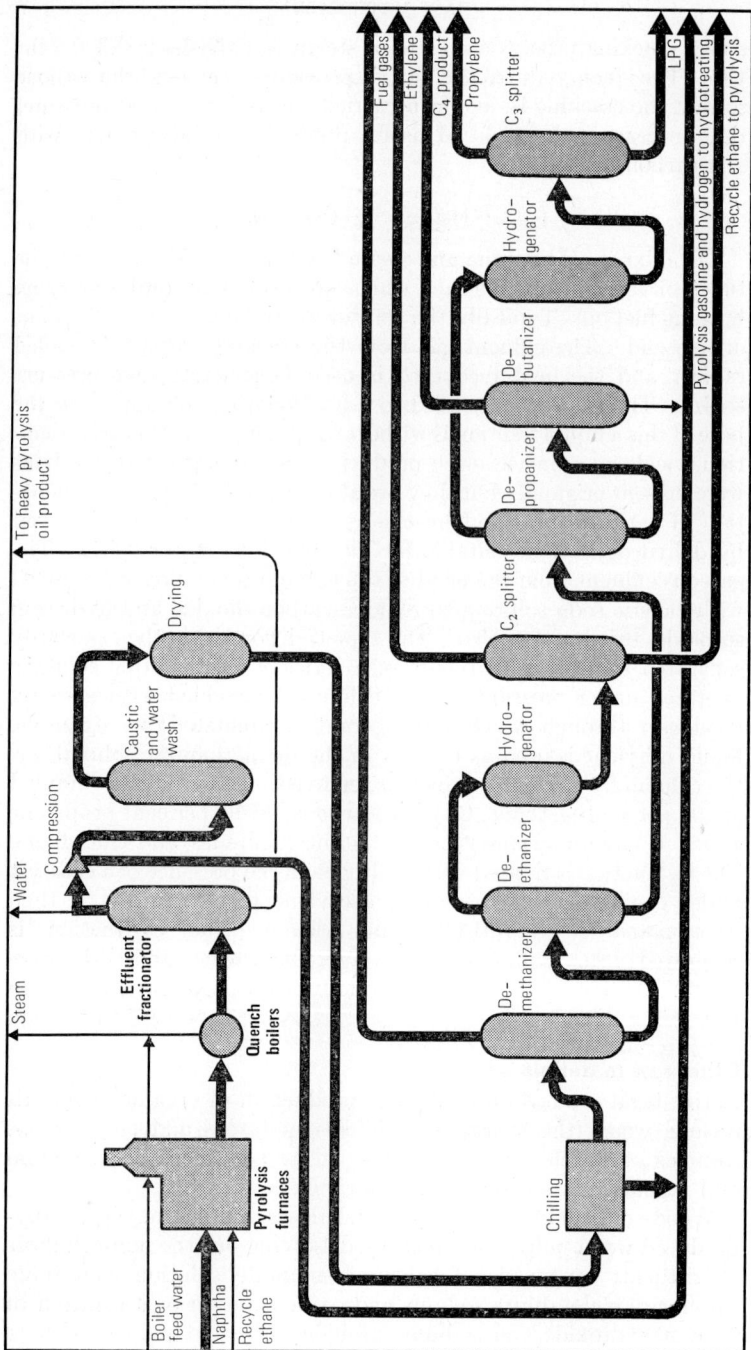

large quantities for the production of paper, viscose rayon, cellophane, cellulose di- and tri-acetate fibres, cellulose di-acetate plastics, nitrocellulose (for explosives, lacquers and celluloid) and cellulose ethers (methyl and ethyl celluloses and sodium carboxymethyl cellulose). Production of cellulosic plastics (largely cellulose acetate plastics) in the UK in 1969 amounted to 10 600 t. The chief object of hard wood distillation, carried out at two plants in the UK (one of which is modern), is the production of wood charcoal but useful by-products are also formed (p 57): certain species of pines are sources of rosin, turpentine and pine oils.

Potatoes and cereal crops, such as maize, rice and wheat, are sources of starch which is used for the production of glucose and glucose syrup by acid hydrolysis. Sucrose produced from sugar cane and sugar beet is used almost entirely as a foodstuff. Beet and sugar cane molasses (*i.e.* the non-crystallizable residue which remains after the extraction of sugar) containing 50–54 per cent of sugars, are used for the production of industrial alcohol. The use of molasses for this purpose has declined considerably owing to competition from industrial alcohol derived from petroleum. Molasses is still, however, of value as a nutrient for the growth of bacteria and mould in the production of lactic acid and citric acid respectively. Corn cobs and sugar cane bagasse (*i.e.* the cane residues remaining after the extraction of sugar) are used as sources of the heterocyclic compound furfural, by digestion with acid under steam pressure.

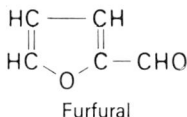

$$HC \overset{\displaystyle |\!|}{\underset{\displaystyle HC}{\quad\quad}} \overset{CH}{\underset{C-CHO}{|\!|}}$$

Furfural

Seaweeds are used as sources of alginic acid and alginates (from *Laminaria cloustoni* and *Ascophylum nodosum*) and agar (from certain species of red seaweed). Apple pomace (the residue from the processing of apples in cider manufacture) and citrus fruit wastes are sources of pectin.

Animal and vegetable fats, although used mainly as foodstuffs and for the production of soap and glycerol, are also used as sources of higher fatty acids. The latter are important in the form of their insoluble metallic salts (*e.g.* zinc stearate) and as raw materials for the manufacture of surface-active agents (p 33). Hides or skins unsuitable for the production of leather and also hide wastes are used in the production of glue. The skins of young animals and also

---

Fig. 3. (Facing page) Ethylene plant. (By courtesy of Kellogg International Corporation, London.)

bones are used to produce gelatine. Milk is the commercial source of the protein casein and the only commercial source of the sugar lactose.

References

1. *The Petroleum Handbook*, 5th edn. London: Shell International, 1966. *Modern petroleum technology*, 3rd edn. London: Inst. of Petroleum, 1962.
2. P. W. Sherwood, *Petrochemical profits for tomorrow: the markets, the technology*. New York: Palmer Publishing Inc., 1966.
3. R. A. Duckworth, 'Ethylene', *Chem. Process Engng*, 1968, **49**(2), 67.

2. High Polymers and Surface-active Agents

High polymers are substances with very high molecular weights made up from units of simple substances by the process of polymerization and they have many uses in industry. Since these materials will be frequently mentioned in the text it is necessary to give a brief account of the chemistry of polymerization and the various types of high polymers available commercially.

Addition polymers

Olefins and their derivatives $CH_2 = CHX$ (vinyl compounds) and $CH_2 = CXY$ (vinylidene derivatives) undergo a process of self-addition under the influence of various catalysts to give products (polymers) whose molecular weights are integral multiples of the molecular weight of the parent compounds (the monomers). The process, known as addition polymerization, may be represented as follows:

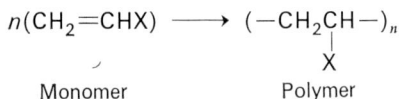

$$n(CH_2{=}CHX) \longrightarrow (-CH_2\underset{\displaystyle X}{\overset{\displaystyle |}{CH}}-)_n$$

Monomer   Polymer

When a polymer is formed from one monomer it is known as a homopolymer. If two different monomers are involved the polymer is known as a co-polymer (p 20). Addition polymerization is started by the addition of a small quantity of an initiator or catalyst to the monomer. Initiators may be substances which decompose into free radicals, the latter initiating polymerization; substances which give rise to cations (cationic initiators); or substances which give rise to anions (anionic initiators).

The most widely used *free radical initiators* are organic peroxides and peroxy esters, benzoyl peroxide being particularly important:

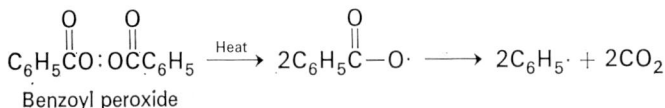

$$C_6H_5\overset{\displaystyle O}{\overset{\displaystyle \|}{C}}O{:}O\overset{\displaystyle O}{\overset{\displaystyle \|}{C}}C_6H_5 \xrightarrow{\text{Heat}} 2C_6H_5\overset{\displaystyle O}{\overset{\displaystyle \|}{C}}{-}O{\cdot} \longrightarrow 2C_6H_5{\cdot} + 2CO_2$$

Benzoyl peroxide

Polymerization of a vinyl derivative through the agency of benzoyl peroxide is assumed to take place as follows, R being a free radical

11

derived from benzoyl peroxide:

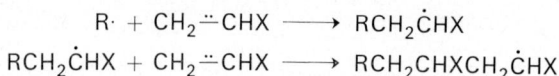

$$R\cdot + CH_2\!=\!CHX \longrightarrow RCH_2\dot{C}HX$$
$$RCH_2\dot{C}HX + CH_2\!=\!CHX \longrightarrow RCH_2CHXCH_2\dot{C}HX$$

and so on, leading eventually to long-chain molecules of the structure $R(CH_2CHX)_nCH_2\dot{C}HX$. Termination of the growth of such chains takes place by union of the ends of two radicals (———$\dot{C}HX$ +$\dot{C}HX$——— $\longrightarrow$ ————$XCH-CHX$———) or by the abstraction of a hydrogen atom by one radical from another. Substances which generate free radicals are the most widely used catalysts for the polymerization of vinyl derivatives. Persulphates are important as initiating agents in the emulsion polymerization of vinyl derivatives (see later). The persulphate ions decompose on heating to give sulphate ion-radicals,

$$S_2O_8^{2-} \xrightarrow{\text{Heat}} 2SO_4^-$$

which initiate polymerization of the vinyl derivative.

Cationic catalysts give rise to carbonium ions in the presence of a monomer and these initiate polymerization. Since carbonium ions are positively charged they are most suitable for polymerizing monomers which contain electron-releasing groups in their molecules. Because of their very high reactivity it is usual to carry out the process in the presence of a diluent, at low temperature. The only cationic catalysts which have been used on the commercial scale are boron trifluoride and anhydrous aluminium chloride, and they are effective only in the presence of water or some other co-catalyst:

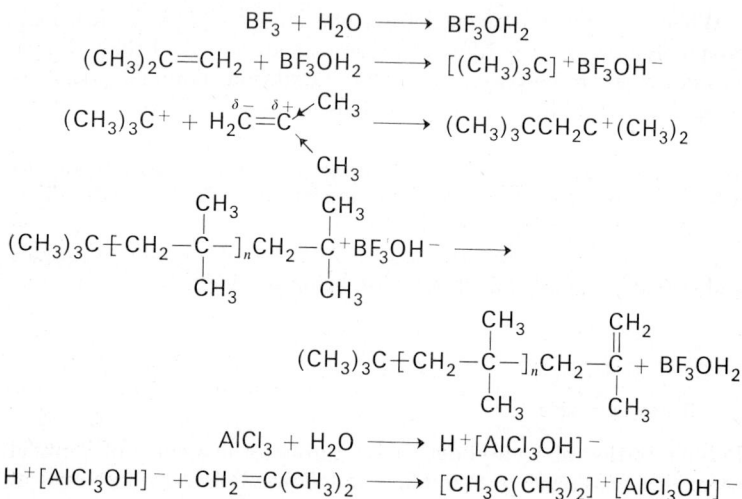

$$BF_3 + H_2O \longrightarrow BF_3OH_2$$
$$(CH_3)_2C\!=\!CH_2 + BF_3OH_2 \longrightarrow [(CH_3)_3C]^+BF_3OH^-$$
$$(CH_3)_3C^+ + H_2\overset{\delta-}{C}\!=\!\overset{\delta+}{C}\!\!\overset{CH_3}{\underset{CH_3}{\diagup\!\!\!\diagdown}} \longrightarrow (CH_3)_3CCH_2C^+(CH_3)_2$$

$$(CH_3)_3C\!\!-\!\![CH_2\!-\!\underset{CH_3}{\overset{CH_3}{\underset{|}{\overset{|}{C}}}}\!-\!]_nCH_2\!-\!\underset{CH_3}{\overset{CH_3}{\underset{|}{\overset{|}{C}}}}{}^+BF_3OH^- \longrightarrow$$

$$(CH_3)_3C\!\!-\!\![CH_2\!-\!\underset{CH_3}{\overset{CH_3}{\underset{|}{\overset{|}{C}}}}\!-\!]_nCH_2\!-\!\underset{CH_3}{\overset{CH_2}{\underset{|}{\overset{||}{C}}}} + BF_3OH_2$$

$$AlCl_3 + H_2O \longrightarrow H^+[AlCl_3OH]^-$$
$$H^+[AlCl_3OH]^- + CH_2\!=\!C(CH_3)_2 \longrightarrow [CH_3C(CH_3)_2]^+[AlCl_3OH]^-$$

The $(CH_3)_3C^+$ cation then adds on to a molecule of isobutene as before. By far the most important application of cationic polymerization is the production of butyl rubber (p 27).

Anionic polymerization, although known for a long time (*e.g.* the polymerization of styrene by NH_2^- ions from sodamide), did not become important until about 1950. Ziegler found that ethylene could be polymerized at about 70 °C and 1.034 MPa (previously it had only been polymerized under very high pressures) in an inert hydrocarbon solvent in the presence of a heterogeneous catalyst. This catalyst was formed by reaction between an aluminium trialkyl and a group 4–6 transition metal compound (*e.g.* titanium tetrachloride, which is now widely used). Polymers produced by this process have only slightly branched molecules and consequently have a higher degree of crystallinity (*i.e.* the polymer chains are arranged in a highly uniform pattern) and therefore a higher m.p. (120–128 °C) than polyethylene produced by the high pressure process:

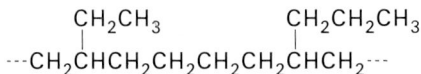

$$\begin{array}{ccc} CH_2CH_3 & & CH_2CH_2CH_3 \\ | & & | \\ \cdots CH_2CHCH_2CH_2CH_2CH_2CHCH_2 & \cdots \end{array}$$

Branched molecular chain in high pressure polyethylene

The polymers produced by the Ziegler process, and by the Phillips process (*see* later), also have a higher S.G. (0.945–0.946) and tensile strength than the high pressure product (S.G. = 0.92, known as low density polyethylene) since the polymer chains, being only slightly branched, are able to pack more closely together.

Polyethylene is produced in the UK by the ICI high pressure process and by the Ziegler and Phillips processes. The ICI process is carried out at a pressure greater than 152 GPa and at about 200 °C in the presence of a trace of oxygen as catalyst. It is particularly important that the feedstock should be free from acetylene.

The Ziegler process has been operated in the UK since about 1958. Ethylene is passed into a hydrocarbon solvent containing a suspension of a catalyst (actually a co-ordination complex) formed by reaction between an aluminium trialkyl (often the triethyl derivative) and titanium tetrachloride at 50–75 °C and 0.5065–0.7091 MPa. The resulting polymer–solvent slurry is then treated with water or dilute acid to decompose the catalyst, and the polymer recovered by filtration.

In the Phillips process (*Fig.* 5) a solid catalyst, such as chromium oxide on a silica–alumina support, and a solvent (cyclohexane) are present in the reactor as a slurry, and the polymer is maintained in solution until after the catalyst is removed from the reaction mixture. The reactor is maintained at 92–149 °C and 0.7091–3.343 MPa. Sulphur-free ethylene is treated to remove water, oxygen and carbon

FIG. 4.   Part of a polyethylene plant based on the ICI process. (By courtesy of Simon–Carves Chemical Engineering Ltd.)

FIG. 5. (Facing page) Production of polyethylene—Phillips Petroleum Co. (Reproduced from the November 1969 issue of *Hydrocarbon Processing*, p 222, by permission of the Gulf Publishing Company, Houston, Texas.)

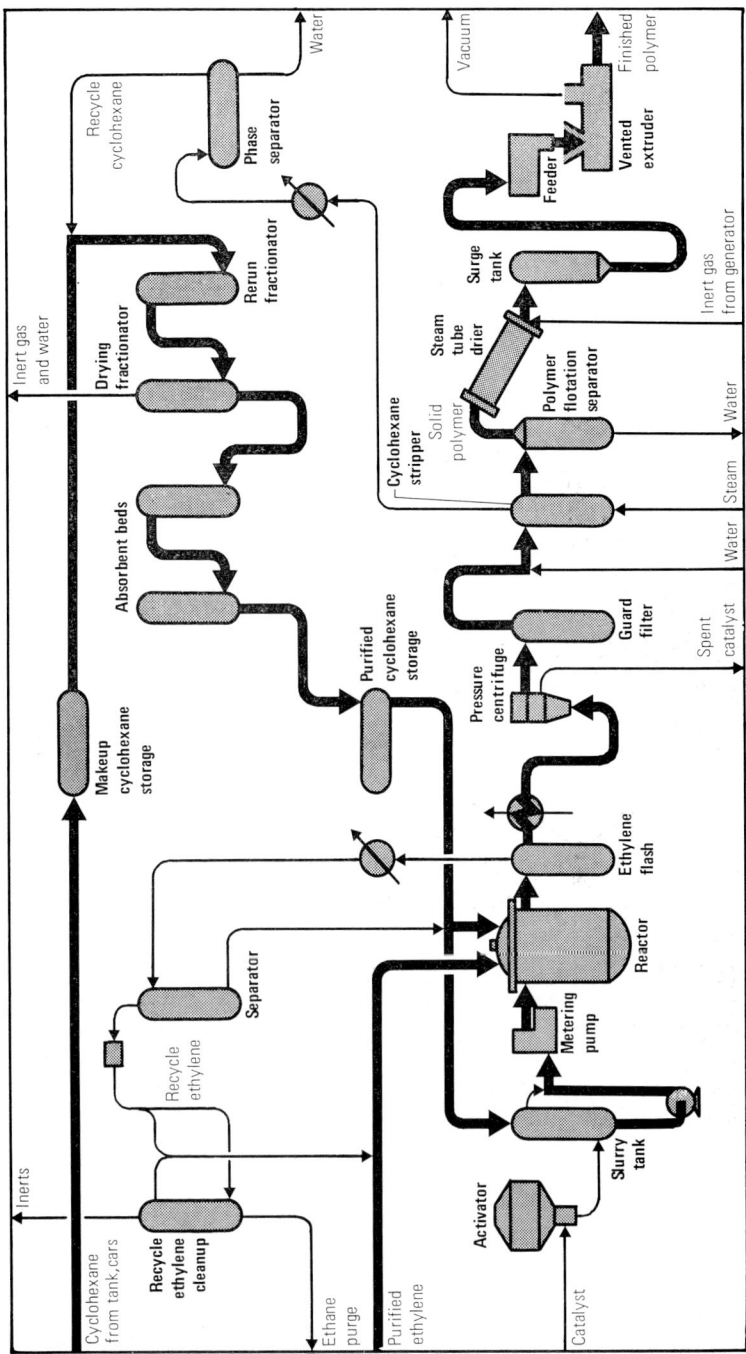

Recycle cyclohexane

Water

Phase separator

Rerun fractionator

Drying fractionator

Inert gas and water

Absorbent beds

Makeup cyclohexane storage

Purified cyclohexane storage

Cyclohexane stripper

Solid polymer

Steam tube drier

Surge tank

Vacuum

Feeder

Vented extruder

Finished polymer

Inert gas from generator

Polymer flotation separator

Water

Steam

Water

Spent catalyst

Guard filter

Pressure centrifuge

Ethylene flash

Reactor

Separator

Recycle ethylene

Metering pump

Slurry tank

Activator

Recycle ethylene cleanup

Ethane purge

Purified ethylene

Catalyst

Cyclohexane from tank, cars

Inerts

dioxide which act as catalyst poisons. The treated ethylene and solvent are charged to the reactor at rates which will give the equivalent of about five weight per cent ethylene. The catalyst is maintained at a concentration of 0.5 per cent or less (based on the weight of solvent). The cyclohexane, in addition to its solvent function, protects the growing polymer chain from chain-breakers, controls the viscosity of the solution and the rate of ethylene consumption so as to ensure good polymer growth, and also dissipates the heat of reaction. Some of the excess solvent and ethylene present in the reaction effluent is flashed off and the polymer solution is then centrifuged and filtered. The polymer is precipitated from the filtrate in a stripper, separated from the resulting slurry, dried and pelleted.

Prior to 1954 high molecular weight polymers had never been prepared from propylene by the use of a substance giving rise to free radicals. The reason for this is the tendency for the radical chains to break down to give olefins and smaller radicals rather than continuing to grow. Then G. Natta and his co-workers in Italy using Ziegler-type catalysts under closely controlled physical conditions succeeded in obtaining high molecular weight polymers from propylene. Previously, the polymers which had been obtained from substituted olefins had a random arrangement of the polymer units. Such polymers are known as *atactic* polymers. In atactic polypropylene the methyl groups are arranged at random relative to each other and the main direction of the polymer chain ($R = CH_3$)

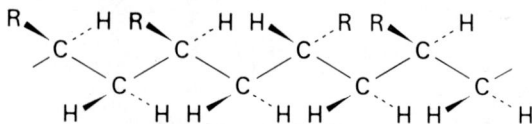

(a bold line denotes a bond projecting out from the plane of the paper towards the observer whilst a broken line denotes a bond directed into the paper). The polypropylene obtained by Natta consisted largely of material having an orderly arrangement of the polymer units, and is said to be *isotactic*.[1-3] In isotactic polypropylene the bonds to the methyl groups all project in the same direction (*i.e.* all the units making up the polymer have the same configuration). Ziegler-type catalysts, on account of their ability

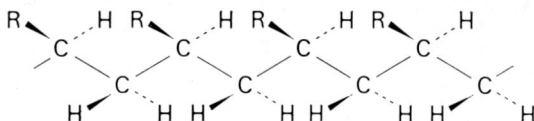

to polymerize propylene to give a stereoregular polymer, are said to be *stereospecific*. It is believed that the stereospecificity of such catalysts is due to the fact that the propylene becomes orientated on the solid surface of the catalyst (*soluble* aluminium–titanium complexes do not effect stereospecific polymerization of α-olefins). Actually, the properties of the polypropylene produced using a Ziegler-type catalyst depend on the composition of the catalyst and on the size and shape of the catalyst particles. Isotactic polypropylene, being highly crystalline, has a high melting point (165–175 °C), a high tensile strength and the lowest density of all commercial plastics (0.906). It is important as a plastic material, for the production of fibres (especially ropes, on account of the high tensile strength) and as packaging film.

Addition polymerizations are carried out by one of four processes. In *mass polymerization* only the monomer(s) and initiator (*e.g.* benzoyl peroxide) are involved. Because of the difficulty of controlling the process it has only been used for polymerizing a few monomers (*e.g.* methyl methacrylate, vinyl ethers and, recently, vinyl chloride). In *solution polymerization* an inert diluent in which monomer(s) and polymer are soluble is employed. The initiator, if it is one which gives rise to free radicals, is also soluble in the solvent. The process is particularly useful for the production of surface-coating compositions, but probably the most important example of its use is the production of polyolefins and butyl rubber (p 27). In *suspension* and *emulsion polymerization*, the monomer is polymerized in the form of a stabilized aqueous suspension or as an emulsion. In the former process the most efficient catalysts are those soluble in the monomer (such as benzoyl peroxide), whereas in emulsion polymerization water-soluble catalysts are used, usually potassium persulphate. Both methods lead to polymers with higher average molecular weights than are obtained by the mass and solution methods. Emulsion polymerization is particularly important, giving a product with a higher average molecular weight than is obtained by the suspension method. It is used for the polymerization of several vinyl derivatives and for the production of synthetic rubbers by the co-polymerization of butadiene and styrene or acrylonitrile.

Table 1 gives some details of the more important addition polymers which are manufactured and used as plastics (polyethylene and polypropylene are also used as fibres). Plastics produced by addition polymerization are known as *thermoplastics* since they soften on heating and harden again on cooling and these processes can be repeated more or less indefinitely. In thermoplastics the molecular chains are held together by weak intermolecular or van der Waals' forces (there are no chemical cross-links) which are overcome by a

Table 1. Addition polymers used as plastics

Monomer	Method of polymerization	Initiator	Polymer	Properties and applications
$CH_2=CH_2$	Mass	O_2	Low density polyethylene	The polyethylenes[4] are wax-like, tough materials with very good dielectric properties and excellent resistance to chemicals. The high-density types have higher tensile strengths and higher softening points than the low-density type. Used for the production of a wide variety of household and other articles (by injection moulding, blow moulding and extrusion), as packaging film* and sheet (this application probably accounts for about one-third of the production of polyethylenes), as piping in the building industry, in the electrical industry and for making fibres
	Solution	$Al(C_2H_5)_3$–$TiCl_4$	High density polyethylene	
	Solution	Cr_2O_3–SiO_2–Al_2O_3	High density polyethylene	*Low density polyethylene accounts for nearly 60 per cent of the plastics used for packaging in the UK
$CH_3CH=CH_2$	Solution	AlR_3–$TiCl_3$	Polypropylene[5,6]	Has a higher softening point than the polyethylenes and is also stiffer and has a harder surface. Used for the production of a variety of articles by injection moulding (car components, household articles, etc.), ropes, woven backing for carpets (the production of fibres probably accounts for about one-third of the polypropylene production), and as film. In the US in 1969, 100 Mlb of polypropylene was used in the motor car industry
$C_6H_5CH=CH_2$	Solution and suspension	Organic peroxide	Polystyrene	Strong but somewhat brittle material with very good dielectric properties and excellent resistance to chemicals. Used for making toys, household articles, refrigerator parts, electrical components and as clear sheet. Also available as toughened polystyrene (a blend of polystyrene with synthetic rubber used, for example, in refrigerators),[7] and as expanded or cellular polystyrene which is widely used as a packaging material and for heat insulation[8]

Table 1—continued

Monomer	Method of polymerization	Initiator	Polymer	Properties and applications
$CH_2{=}CHCl$	Mass Suspension Emulsion	Not known Monomer-soluble such as an organic peroxide $K_2S_2O_8$	Polyvinyl chloride[9,10]	Available as rigid (unplasticized) polyvinyl chloride, as the plasticized material (see later), and as paste. The rigid material (which accounts for 17–20 per cent of the production of polyvinyl chloride) is used in the building industry in the form of pipes, guttering, *etc.*, for making trays, tank linings, fume cupboards, helmets, *etc.* Plasticized polyvinyl chloride is used for making floor tiles ('vinyl' tiles), as a leather substitute, for making hose pipe, raincoats, conveyor belting, for cable coating, as film, and for making a variety of household articles. In order to minimize deterioration of polyvinyl chloride when exposed to heat or light, stabilizers are incorporated
$CH_2{=}C(CH_3)$ $\quad COOCH_3$	Mass Solution Emulsion	Benzoyl peroxide Monomer-soluble substance $K_2S_2O_8$	Polymethyl-methacrylate[11]	Tough transparent material softening at about 100 °C, important in sheet form as a glass substitute, compared with which it has a higher light transmission. Biaxially stretched polymethyl methacrylate sheet is important, because of its much higher impact strength, in the motor and aircraft industries. Polymethyl methacrylate is also used for making optical parts and various moulded articles such as outdoor signs, car components, protective goggles and household goods
$CF_2{=}CF_2$	Emulsion	$(NH_4)_2S_2O_8$	Polytetra-fluoroethylene[12] (PTFE)	Produced on a relatively small scale and is expensive. It is highly crystalline, has a high softening point (327 °C), outstanding resistance to chemicals, excellent heat resistance, very good dielectric properties, and has the lowest coefficient of friction of any known substance. Because of its high softening point and lack of flow it has to be fabricated by the methods of powder metallurgy. Used for cable insulation, in chemical plant, and for coating frying pans and bearings

small input of heat. The degree of cohesion between the chains depends on the nature of the functional groups present in the chains. Thus, polyvinyl chloride is much harder than polyvinyl acetate because the acetate groups are bulky and prevent the chains coming as near to each other as they do in the case of polyvinyl chloride.

Other important addition polymers include polyalkyl acrylates and polyalkyl vinyl ethers.

If vinyl chloride is polymerized in the presence of another monomer such as vinyl acetate, the product which results is known as a *co-polymer*, the vinyl acetate units being distributed along the co-polymer chains at random:

$$\cdots CH_2-CH-CH_2-CH-CH_2-CH-CH_2-CH-CH_2-CH\cdots$$
$$\quad\quad\ |\quad\quad\quad\ |\quad\quad\quad\ |\quad\quad\quad\ |\quad\quad\quad\ |$$
$$\quad\quad\ Cl\quad\quad\ OCOCH_3\ \ Cl\quad\quad\quad Cl\quad\quad\ OCOCH_3$$

Polyvinyl chloride is a hard substance, whereas polyvinyl acetate is a relatively low-melting resin and hence is not used as a plastic material. In the case of the co-polymer the vinyl acetate units have the effect of producing a softer product than polyvinyl chloride, the softening process being known as *internal plasticization*. The properties of the co-polymer vary according to the composition of the monomer mixture employed. Important co-polymers are also produced from butadiene and styrene, butadiene and acrylonitrile, and styrene and acrylonitrile. Nearly all commercial co-polymers are of the random co-polymer type. When the monomer units are present in the co-polymer chains in blocks, *e.g.* AAAAAABBBBBBAABBBB, the co-polymer is referred to as a *block co-polymer*. It is also possible for units of a second monomer to be grafted on to a polymer chain,

```
A A A A A A A A A A A
  B       B       B
  B       B       B        Graft co-polymer
  B       B       B
  B       B
          B
```

the co-polymer being then referred to as a *graft co-polymer*. ABS plastics (p 88) are of this type.

Additional materials called *plasticizers* are often added to thermoplastic compositions. These are low-volatility substances (usually liquids) incorporated in amounts of from 20–50 per cent into thermoplastic moulding compositions, polyvinyl chloride pastes, surface-coating compositions and adhesives. Their function in moulding compositions is to give satisfactory flow properties, thus facilitating moulding, extrusion and calendering, and enabling processing to take place at temperatures below the decomposition temperature of the thermoplastic. They also confer on the plastic material greater flexibility and extensibility and, in some cases, may also reduce flammability. The most important substance plasticized is polyvinyl chloride, together with cellulose acetate and polyvinyl acetate. Plasticizers are divided into two classes: simple plasticizers which have relatively low molecular weights, and resinous or polymeric plasticizers which have much higher molecular weights. The most important plasticizers of the former class are the esters of dicarboxylic acids and phosphoric acid, including the dialkyl phthalates, adipates and sebacates from long-chain alcohols [*e.g.* di-2-ethylhexyl adipate and triaryl phosphates (tricresyl phosphate is particularly important)]. A typical polymeric plasticizer is the linear polyester from polypropylene glycol and a dicarboxylic acid such as adipic or sebacic acid.

Condensation polymers

These are formed by a series of condensation reactions, *i.e.* unions between two or more molecules with elimination of a simple molecule such as water, an alcohol, hydrogen chloride, ammonia, *etc.*

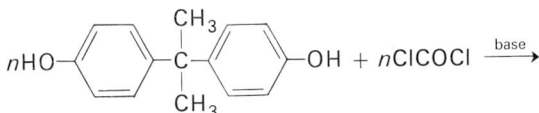

2,2-bis-*p*-hydroxyphenylpropane

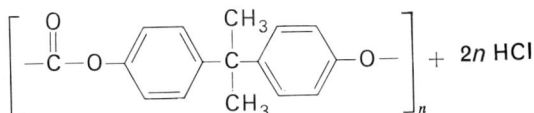

polycarbonate

Dimethyl terephthalate

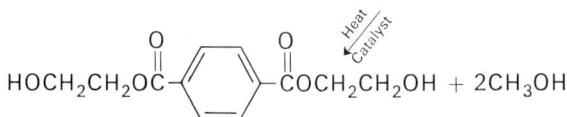

Dihydroxydiethylterephthalate

$$n[HOCH_2CH_2O\overset{O}{\overset{\|}{C}}\!\!\!-\!\!\!\langle\!\!\!\bigcirc\!\!\!\rangle\!\!\!-\!\!\!\overset{O}{\overset{\|}{C}}OCH_2CH_2OH] \xrightarrow{\text{Heat}}$$

$$HOCH_2(CH_2O\overset{O}{\overset{\|}{C}}\!\!\!-\!\!\!\langle\!\!\!\bigcirc\!\!\!\rangle\!\!\!-\!\!\!\overset{O}{\overset{\|}{C}}OCH_2)_nCH_2OH + (n-1)HOCH_2CH_2O$$

Polyethyleneterephthalate (Terylene)

$$n[^-O_2C(CH_2)_4CO_2^- H_3\overset{+}{N}(CH_2)_6\overset{+}{N}H_3]$$

Nylon salt from adipic acid
and hexamethylene diamine $+ \ ^-O_2C(CH_2)_4CO_2^- H_3\overset{+}{N}(CH_2)_6\overset{+}{N}H_3$

$$^-O_2C(CH_2)_4CO[NH(CH_2)_6NHCO(CH_2)_4CO]_nNH(CH_2)_6\overset{+}{N}H_3 + nH_2O$$

Nylon 66 (a polyamide)

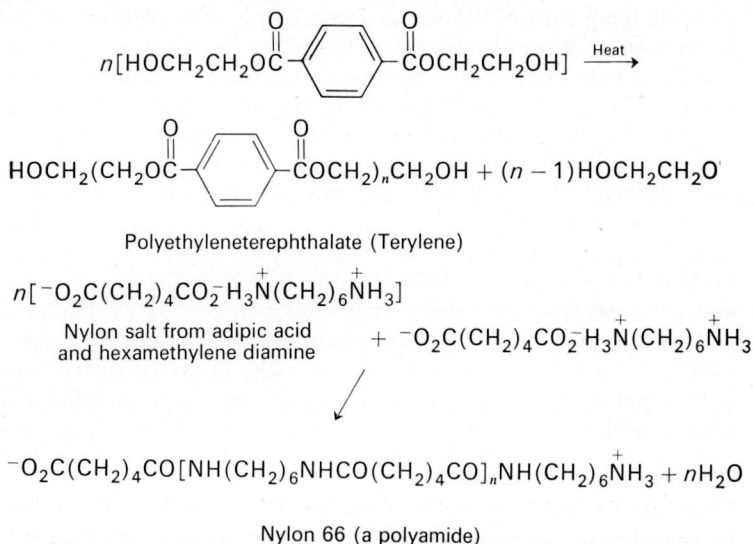

When the reactants which give rise to condensation polymers are difunctional, *i.e.* each has not more than two reactive groups (*e.g.* a dicarboxylic acid and a dihydric alcohol, or a diamine and a dicarboxylic acid), the resulting polymer has a linear structure and is thermoplastic. Important examples of thermoplastics made by condensation polymerization are nylon 6 (polycaprolactam), nylon 66 and polycarbonates from diphenols, such as diphenylolpropane, and phosgene. The process also gives rise to some of the most important synthetic fibres, such as the nylons (polyamides) and the polyester, polyethyleneterephthalate (Terylene). The nylons are, of course, most important as fibres but are also used as plastics because of their high mechanical strength and resistance to abrasion, *e.g.* for gear wheels.

If one of the reactants in a polycondensation is trifunctional (*i.e.* has three functional groups, such as glycerol) then branched-chain molecules and also cross-linking between molecular chains can occur. Thus, low molecular weight condensation polymers formed by the catalyzed reaction between phenol and formaldehyde, such as the following,

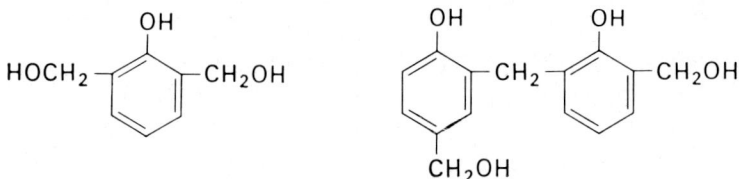

$$HOCH_2\!\!-\!\!\langle\!\!\!\bigcirc\!\!\!\rangle\!\!\!-\!\!\!CH_2OH$$
(with OH above ring)

$$\langle\!\!\!\bigcirc\!\!\!\rangle\!\!\!-\!\!\!CH_2\!\!-\!\!\langle\!\!\!\bigcirc\!\!\!\rangle\!\!\!-\!\!\!CH_2OH$$
(with OH above each ring and CH_2OH below)

can undergo further condensation reactions and also cross-linking:

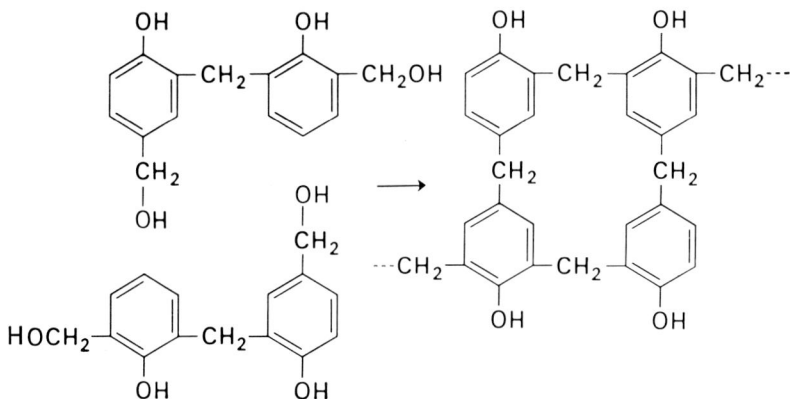

The greater the number of —CH$_2$— cross-links, the greater the rigidity of the product and the lower its solubility in solvents. The product is known as a *thermosetting* resin since once it has been formed it cannot again be softened. Considerably more heat energy would be required to break the cross-links (which would result in decomposition of the resin) than is necessary to melt thermoplastics.

The cross-links may also be produced by a third substance. Thus, the linear unsaturated polyester formed by the reaction between a mixture of maleic anhydride, phthalic anhydride and propylene glycol is cross-linked by means of styrene to give a thermosetting product (the thermosetting process is carried out in the presence of glass fibres).

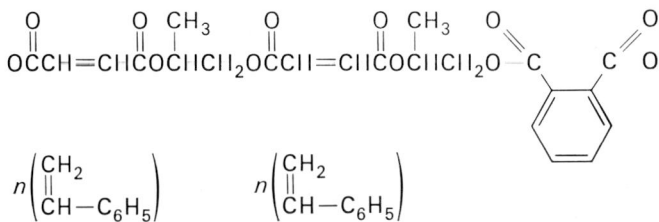

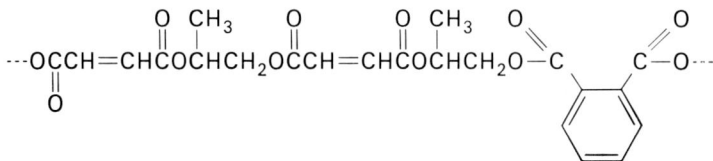

$$
\begin{array}{c}
\text{O} \quad \vdots \quad \text{O} \quad \text{CH}_3 \quad\quad \text{O} \quad \vdots \quad \text{O} \quad \text{CH}_3 \quad\quad \text{O} \quad\quad\quad\quad \text{O}\\
\text{---OCCHCHCOCHCH}_2\text{OCCHCHCOCHCH}_2\text{O—C} \qquad \text{C—O---}\\
\begin{pmatrix} \text{CH}_2 \\ \text{CH—C}_6\text{H}_5 \end{pmatrix}_n \quad\quad \begin{pmatrix} \text{CH}_2 \\ \text{CH—C}_6\text{H}_5 \end{pmatrix}_n\\
\text{O} \qquad\qquad\qquad \text{O} \qquad\qquad\qquad \text{O} \qquad\qquad \text{O}\\
\text{---OCCHCHCOCHCH}_2\text{OCCHCHCOCHCH}_2\text{O—C} \qquad \text{C—O---}\\
\vdots \quad \text{O} \quad \text{CH}_3 \qquad\qquad \vdots \quad \text{O} \quad \text{CH}_3
\end{array}
$$

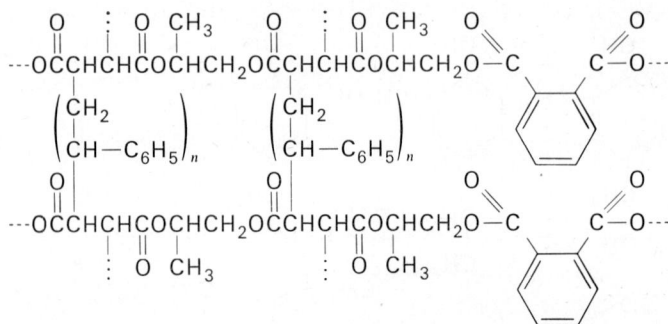

Table 2 lists the more important thermosetting polymers and the raw materials used for their production.

Synthetic fibres[21]

The ability to form fibres with useful properties is possessed by *linear* polymers with a sufficiently high molecular weight. For example, for nylon 66 (p 22) the minimum molecular weight is about 22 000. With molecular weight increasing above the optimum value desirable for the formation of fibres with satisfactory properties, the solubility in solvents decreases and the softening point increases. The solution properties decide whether the polymer can be spun from solution, whilst if the softening point is too high it is not possible to melt spin the polymer. In the production of a polymer suitable for conversion to fibres the process of polymerization is stopped by addition of a small amount of a monofunctional constituent to the reaction mixture. Fibres are produced from the polymers by melt spinning, *i.e.* forcing the molten polymer (*e.g.* nylon or Terylene) through spinnerets, or by dissolving the polymer in a solvent and forcing the solution through spinnerets into warm air to remove the solvent ('dry' spinning, as for cellulose acetates) or by forcing a solution of the polymer through a spinneret into a regenerating solution which causes precipitation of the fibres ('wet' spinning, as for viscose rayon). After spinning, the fibres may be stretched to increase their strength. The linear molecular chains which make up fibres are oriented parallel to the fibre axis and to each other in varying degrees, and stretching increases the degree of orientation. The production of polyester fibres has been surveyed.[22]

Table 3 lists the more important man-made fibres and the UK production statistics.

Table 2. Thermosetting plastics

Polymer	Raw materials	Applications
Phenolic resins [13]	Phenol and/or cresols and formaldehyde	Used for the production of moulding powders, the properties of the mouldings (produced by compression moulding) depending on the type of filler used (wood flour, asbestos, cotton, *etc.*). Phenolic resins are also used as adhesives, heat-hardening varnishes, and for making plywood and laminated plastics
Urea–formaldehyde resins Melamine–formaldehyde resins Both types are known as amino-resins [14,15]	Urea and formaldehyde Melamine and formaldehyde	The amino-resins are used for the production of moulding powders and laminated plastics, crease-resistant finishes for textiles, for strengthening paper, as adhesives, and in a modified form for the production of stoving finishes. Melamine–formaldehyde resins are particularly important in the form of quality tableware
Epoxy resins [16]	Resins from 2,2-bis-*p*-hydroxyphenylpropane (p 124) and epichlorohydrin (p 55) plus curing agents	Used in surface-coating compositions and for their excellent electrical insulation properties, and as adhesives between wood and metal
Glass-fibre reinforced polyester resins [17]	Polyester from maleic anhydride, phthalic anhydride and propylene glycol, cross-linked with styrene	These possess a very high strength-to-weight ratio and have the added advantage that the thermosetting process need not be carried out under pressure. They have become important in the building industry, for boat-building, and also have other applications
Modified alkyd resins [18]	Phthalic anhydride, glycerol and/or pentaerythritol and fatty acids	Used for the production of hard gloss paints and stoving paints
Flexible and rigid polyurethanes [19,20]	Di-isocyanates, particularly tolylene di-isocyanates, and either a polyester or the polyether from glycerol and propylene oxide	The two types differ in the degree of cross-linking. The foamed structure is produced *in situ* by the action of water on the polymer with liberation of carbon dioxide. The flexible forms are used as rubber substitutes. The rigid types are used for heat and sound insulation. Polyurethanes are also used as surface-coating materials and to a limited extent as rubbers

Total production of thermosetting resins in the UK in 1969 was 320 700 t

Table 3. UK man-made fibres production in 1968

Fibre	Mlb
Polyamides (*i.e.* nylons 6 and 66)	300
Polyester (Terylene)	148
Acrylic	130
Others	14
Viscose—continuous filament	127
Viscose—staple fibre	366
Acetate—continuous filament	88
Acetate—staple fibre	29

The nylons, polyester and acrylic fibres account for more than 95 per cent of the production of synthetic fibres. Total production of synthetic fibres in the UK in 1969 was 260 400 t. World production of man-made fibres during 1969 was 8 Mlb and of synthetic fibres was 2.2 Mlb.

Synthetic rubbers

Natural rubber is a polymer of isoprene ($CH_2 = CMeCH = CH_2$).

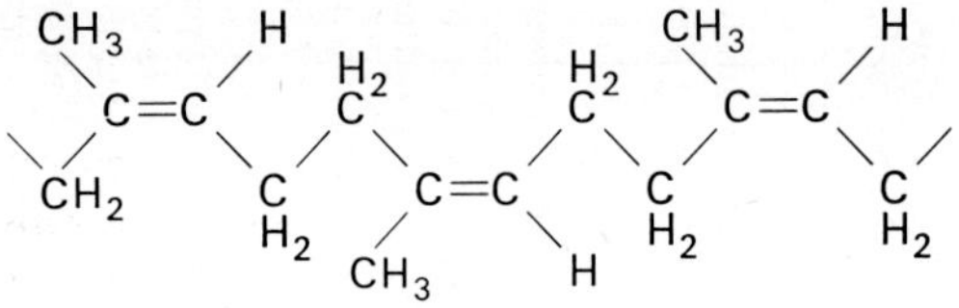

Natural rubber

It exhibits no crystallinity as the molecular chains are coiled and tangled. However, under tension, the chains straighten out and a sharp x-ray diffraction pattern is obtained, indicating crystallinity. For nearly all applications the properties of natural rubber are unsatisfactory and it is necessary to modify it by the process of vulcanization, *i.e.* by heating the rubber with a small amount of sulphur which cross-links the structure.

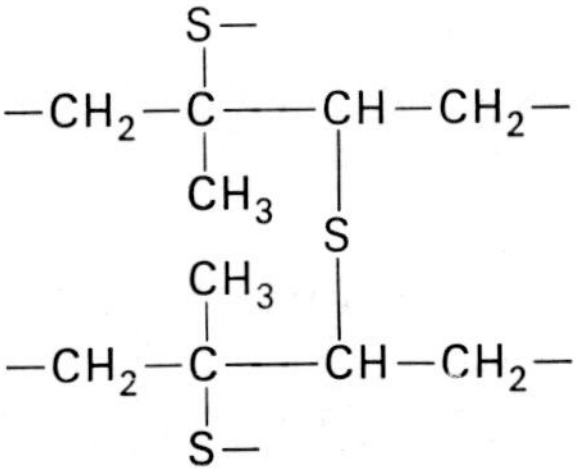

Vulcanized rubber

Since only a few per cent of sulphur is used only a small number of cross-links are formed and, consequently, vulcanized rubber is far less rigid than thermosetting resins (p 23). In addition to sulphur, other substances are incorporated into the rubber mix, *viz.* vulcanization accelerators (organic substances), zinc oxide (to increase the efficiency of the accelerators), reinforcing agents (carbon black is by far the most important of these), fillers, a plasticizer to facilitate processing of the rubber mix, an antioxidant and coloured pigment.[23,24] Natural rubber accounts for about 40 per cent of the world market for elastomers. World consumption of natural and synthetic rubbers in 1970 was 6.8 Mt.

The most important synthetic rubber in terms of quantity produced is the styrene–butadiene co-polymer rubber (SB-R), made by co-polymerizing butadiene and styrene (70–75 : 30–25) in aqueous emulsion. World production of SB-R in 1969 has been estimated as 3.24 Mt (cf. natural rubber, 2.87 Mt). SB-R is a general purpose rubber (i.e. it may be used for most of the applications of natural rubber) and is used to partially replace natural rubber in tyres. Figure 6 illustrates the production of SB-R. Polymerization is stopped (Stage 6) when about 60 per cent of the reactants have co-polymerized.

Butyl rubber production (Fig. 7) represents by far the most important application of cationic polymerization. Liquid isobutene and a small quantity of isoprene (see also Table 4) are diluted with methyl chloride and fed to reactors to which aluminium chloride is added continuously. The co-polymerization takes place at about $-60\,°C$ (propylene and ethylene refrigeration circuits are essential ancillaries of the plant). A slurry of butyl rubber is produced which overflows to a flash drier and stripper where methyl chloride and unchanged reactants are removed by steam and vacuum stripping. The crumb-like particles of rubber slurry in water are then dried by a series of extruders and finally pressed into bales.

Table 4 lists some of the more important synthetic rubbers produced commercially.[25–27] Production of synthetic rubbers in the UK in 1967 amounted to 200 000 t. World consumption of synthetic rubbers in 1970 was 4.6 Mt.

Other rubbers produced in lesser quantities include silicone rubbers which are expensive but, unlike other rubbers, can be used over a much wider temperature range, and polysulphide rubbers, noted for their outstanding resistance to solvents but only used to a limited extent on account of their odour.

Surface-active agents (other than soaps)[28,29]

These may be divided into anionic, cationic and non-ionic types. All three classes have hydrophobic (water-hating) long hydrocarbon

---

Fig. 6. (Page 28) Production of SB-R. (Reproduced from Not From Trees Alone by courtesy of the British Association of Synthetic Rubber Manufacturers.)

Fig. 7. (Page 29) Butyl rubber production. (By courtesy of Esso Petroleum Co. Ltd.)

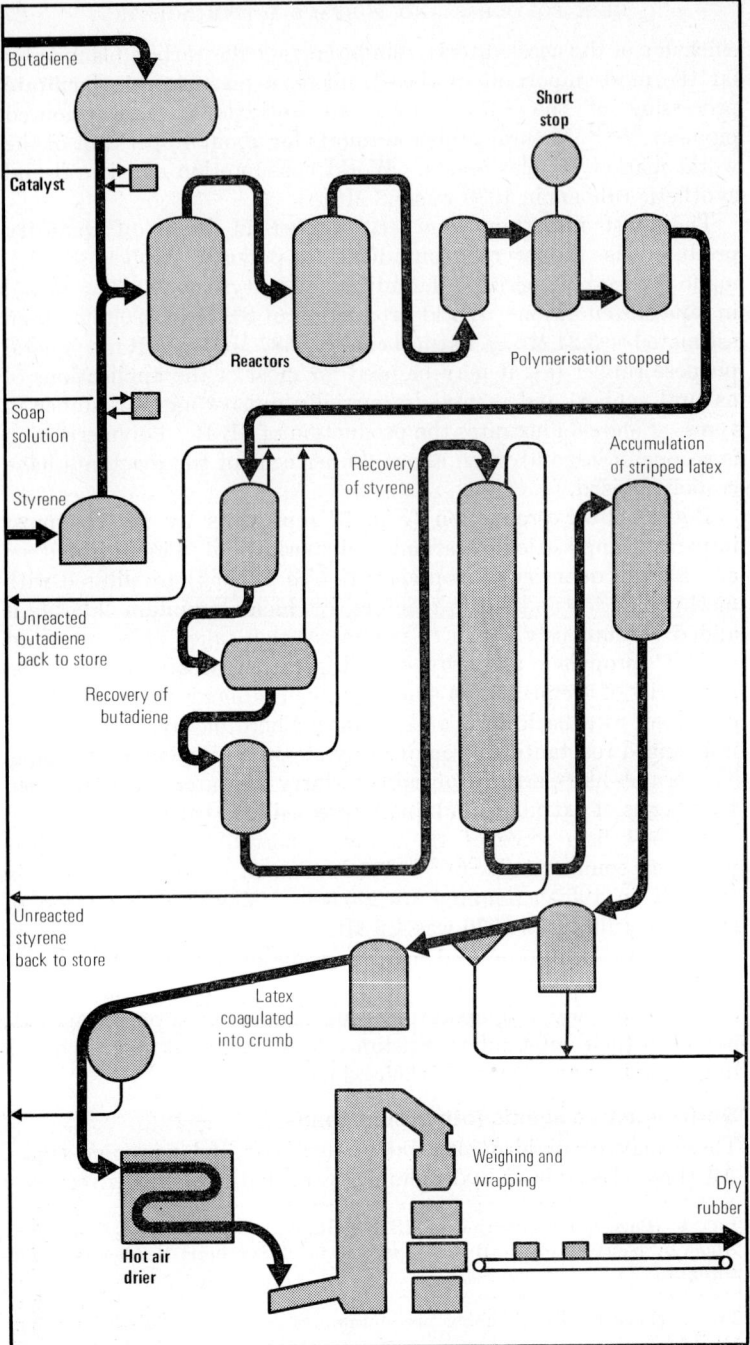

Butadiene

Catalyst

Short stop

Reactors

Polymerisation stopped

Soap solution

Styrene

Recovery of styrene

Accumulation of stripped latex

Unreacted butadiene back to store

Recovery of butadiene

Unreacted styrene back to store

Latex coagulated into crumb

Weighing and wrapping

Dry rubber

Hot air drier

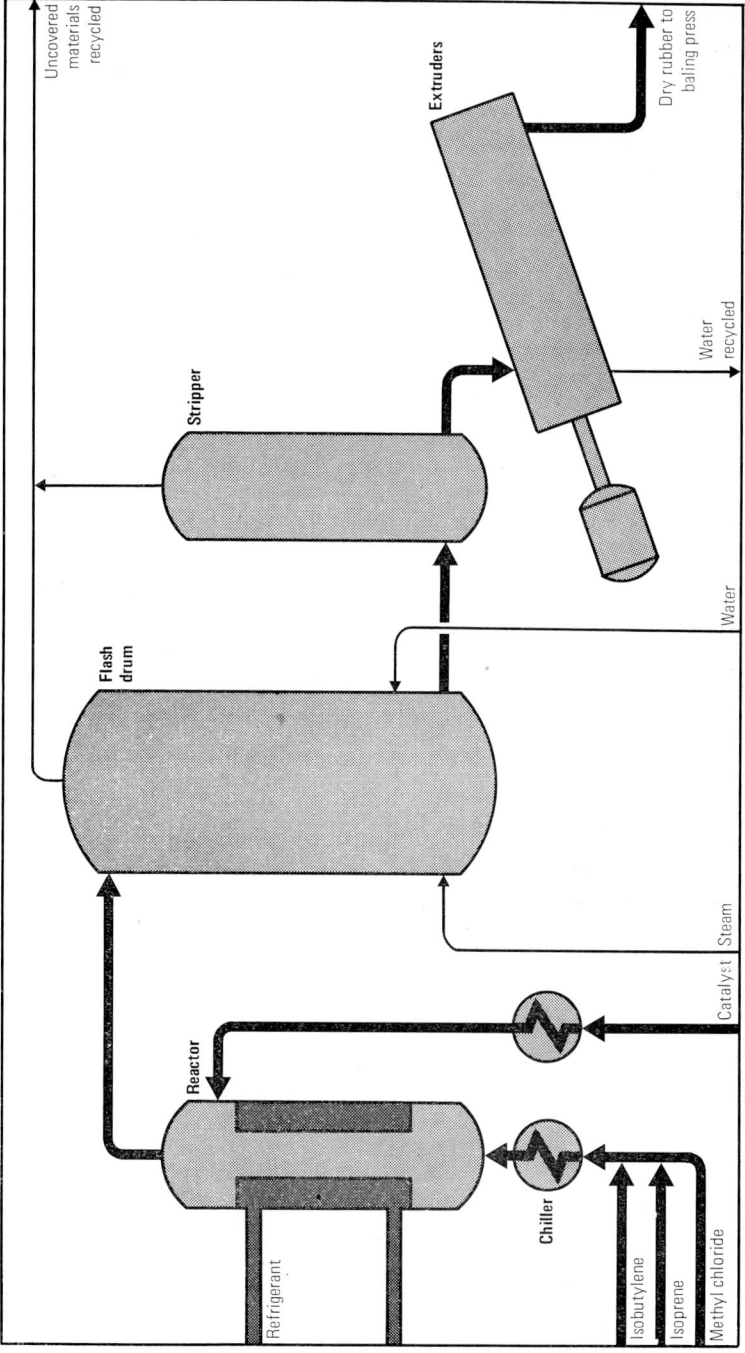

Uncovered materials recycled

Stripper

Extruders

Dry rubber to baling press

Water recycled

Flash drum

Water

Catalyst

Steam

Reactor

Chiller

Refrigerant

Isobutylene

Isoprene

Methyl chloride

chains, solubility being conferred by the formation of anions or cations or by means of a hydroxyl group.

Anionic surface-active agents include sodium alkyl/aryl sulphonates (see later), sodium alkyl sulphates [*e.g.* sodium lauryl sulphate, $CH_3(CH_2)_{10}CH_2OSO_3^-Na^+$] from higher fatty alcohols derived from fats or from ethylene (see later), sodium secondary alkyl sulphates (produced by treating C_{10}–C_{18} α-olefins with sulphuric

Table 4. Synthetic rubbers

Name of rubber	Raw materials and method of polymerization	Properties and applications
SB-R	Butadiene (70–75 per cent) and styrene (30–25 per cent). Emulsion co-polymerization at about 7 °C	Reinforced with carbon black and vulcanized as for natural rubber. Compared with the latter it has better ageing properties and abrasion resistance but is not so resilient, does not resist tearing so well and also has a greater tendency to heat build-up. It is used principally for the production of tyres (about 70 per cent of the rubber used for tyre production consists of SB-R) but also in the footwear industry and for making other rubber goods. Another application is for the production of high impact polystyrene. SB-R latex is also used as an adhesive in the upholstery trade and as a carpet backing
Hard SB-R	Butadiene (10–30 per cent) and styrene (90–70 per cent). Emulsion co-polymerization	Hard rubbers with very good abrasion and tear resistance—used as reinforcing agents for SB-R in the footwear trade, in floor tiles, gaskets, *etc.*
Nitrile rubber	Butadiene (45–80 per cent) and acrylonitrile (55–20 per cent). Emulsion co-polymerization	Has better high temperature characteristics than SB-R and is more resistant to oils and solvents (resistance increases the higher the ratio of acrylonitrile to butadiene, but flexibility at low temperatures also decreases) and also has a high abrasion resistance. It is used for making end-products which have to stand up to oils and solvents, including gaskets, oil seals (*e.g.* in motor cars), hoses, aircraft fuel tanks, *etc.* A blend with polyvinyl chloride has a high impact strength and very high abrasion resistance. It is used chiefly for making soles in the footwear industry

Table 4—continued

Name of rubber	Raw materials and method of polymerization	Properties and applications
High *cis*-polybuta-diene rubber	Butadiene. Solution polymerization in the presence of a Ziegler-type catalyst	Has excellent abrasion resistance when reinforced with carbon black and also higher resilience than natural rubber and good low temperature properties. Blended with natural rubber or SB-R in heavy-duty tyres. Also has non-tyre applications (*e.g.* for the production of ABS polymers, p 88). In the US more than 90 per cent of the poly-butadiene rubber produced is used in tyres. Production in Western Europe in 1970 was expected to be 180 000 t
High *cis*-polyiso-prene rubbers	Isoprene. Solution polymerization in the presence of a Ziegler-type catalyst or lithium butyl	The rubber produced by the use of a Ziegler-type catalyst has a higher *cis*-content than the other. These rubbers can be blended with SB-R or polybutadiene rubber. Their widespread use has been limited by the relatively high cost of isoprene. They are chiefly used for non-tyre applications at present, *e.g.* for shoe soles and heels and mechanical goods. In some applications they are superior to natural rubber. Consumption of polyisoprene rubber in the US in 1970 is expected to be about 400 Mlb
Poly-chloro-prene or -neo-prene	Chlroroprene, $CH_2{=}CHC(Cl)$ $=CH_2$, (from acetylene or butadiene)	Expensive compared with natural rubber or SB-R and has a higher S.G. (1.24). Has a high tensile strength with carbon black as reinforcing agent, exhibits non-flammability, has high resistance to attack by oxygen and ozone, good resistance to oils and solvents and excellent adhesive characteristics. Used for cable sheathing, conveyor belting, gaskets and moulding for cars, adhesives, *etc.* 1970 Western European production was expected to be 75 000 t
Butyl rubber	Isobutene and 2–3 per cent of isoprene, which supplies unsaturation centres in the co-polymer thus enabling it to be vulcanized by sulphur. Solution polymerization in methyl chloride, catalysed by aluminium chloride	Permeability to air is only about $\frac{1}{10}$ that of natural rubber. Tough and resistant to attack by oxygen, ozone and chemicals. Better resistance to heat than natural rubber. Up to 1957 the major use of butyl rubber was for the production of inner tubes but in this year the tubeless tyre was introduced and new outlets have had to be found. Now it is also used as a shock absorber in the automobile industry, as thin sheeting in civil engineering and agriculture and as a lining material for tubeless tyres. The world market for butyl rubber now exceeds 300 000 t/a

Table 4—continued

Name of rubber	Raw materials and method of polymerization	Properties and applications
Ethylene–propylene ter-polymer rubber (EPT)	Ethylene and propylene and 2–20 per cent of a third monomer (such as dicyclopentadiene), the latter providing unsaturation centres to enable the terpolymer to be vulcanized by sulphur. Solution polymerization in the presence of a Ziegler-type catalyst formed by reaction between vanadium tetra-chloride or oxychloride and an aluminium trialkyl	Has a very high resistance to oxygen and ozone, a low S.G. (0.877) and excellent heat-resistance. Competitive with other rubbers in non-tyre applications but also used for making the side-walls of tyres

acid and neutralizing the sulphation mixture), and sodium alkyl

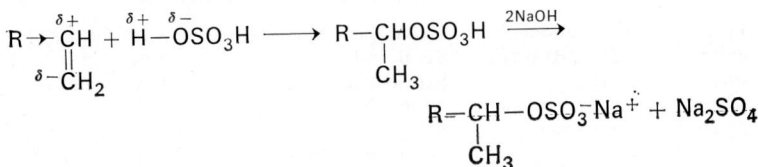

$$R \rightarrow \underset{\underset{\delta-}{\overset{\delta+}{CH}}}{\overset{\delta+}{CH}} + \underset{}{\overset{\delta+}{H}} - \overset{\delta-}{O}SO_3H \longrightarrow R - \underset{\underset{CH_3}{|}}{CHOSO_3H} \xrightarrow{2NaOH}$$

$$R = CH - OSO_3^- Na^+ + Na_2SO_4$$
$$\underset{CH_3}{|}$$

sulphonates, $C_nH_{2n+1}SO_3^-Na^+$, (produced by sulphochlorinating long-chain paraffins and hydrolysing the products with alkali).

Sodium alkyl/aryl sulphonates are produced in greater quantity than any other type of anionic surface-active agent and are the basis of most household washing powders. Production of such detergents is expected to be in excess of 1600 Mlb in the Western hemisphere in 1972. They were formerly made by alkylating benzene with propylene tetramer (produced by polymerizing propylene over a phosphoric acid catalyst on an inert support at 200 °C under pressure), sulphonating, and neutralizing the sulphona-tion mixture. Detergents of this type, with a branched carbon chain

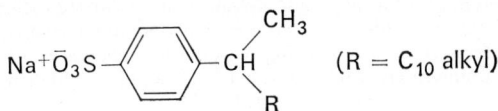

$$Na^+\bar{O}_3S - \bigcirc\!\!\!\!\!\bigcirc - \underset{\underset{R}{|}}{CH} \overset{CH_3}{\diagup} \qquad (R = C_{10} \text{ alkyl})$$

attached to the aromatic ring, are not readily broken down by micro-organisms (*i.e.* they are not *biodegradable*) and gave rise to foaming problems at sewage works and in rivers. They were, therefore, withdrawn by the manufacturers and replaced by products containing an unbranched carbon chain, described as biologically 'soft' or biodegradable detergents (actually such detergents are only 85–95 per cent biodegradable). These may be made by alkylating benzene with chlorinated n-C_{10}–C_{13} paraffins (the n-paraffins are obtained from kerosene and gas oil fractions by the molecular-sieve process described on p 37) in the presence of anhydrous aluminium chloride as catalyst. The product is then sulphonated and the sulphonic acid converted to its sodium salt. Alternatively, 1- or α-olefins can be obtained by cracking wax or by polymerization of ethylene in the presence of a Ziegler catalyst, and may be used to alkylate benzene in the presence of anhydrous aluminium chloride or anhydrous hydrogen fluoride. It should be noted that the sodium salts of long-chain alkyl sulphates (*e.g.* sodium lauryl sulphate) are biodegradable. These are usually made by sulphating fatty alcohols, but they may also be produced by sulphating alcohols, obtained by polymerizing ethylene in the presence of a Ziegler catalyst, oxidizing the product with air to form the alkoxides and hydrolysing the latter with dilute sulphuric acid. This process is used in the US.

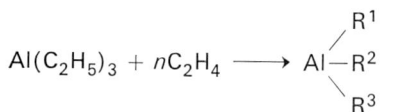

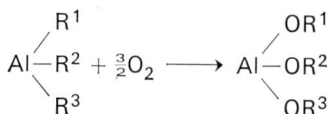

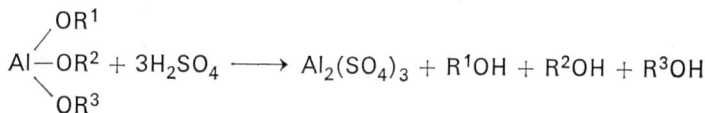

$$Al(C_2H_5)_3 + nC_2H_4 \longrightarrow Al{\Big\langle}{\overset{\displaystyle R^1}{\underset{\displaystyle R^3}{-R^2}}}$$

$$Al{\Big\langle}{\overset{\displaystyle R^1}{\underset{\displaystyle R^3}{-R^2}}} + \tfrac{3}{2}O_2 \longrightarrow Al{\Big\langle}{\overset{\displaystyle OR^1}{\underset{\displaystyle OR^3}{-OR^2}}}$$

$$Al{\Big\langle}{\overset{\displaystyle OR^1}{\underset{\displaystyle OR^3}{-OR^2}}} + 3H_2SO_4 \longrightarrow Al_2(SO_4)_3 + R^1OH + R^2OH + R^3OH$$

Cationic surface-active agents are produced in relatively small quantities compared with those of the other two classes. A typical example is cetyl pyridinium chloride,

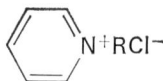

They are not good detergents and are only used in applications where

the anionic types would be unsuitable, particularly in industrial processes such as in the textile industry and in the food processing industry where their germicidal properties are of value. World production of cationic surface-active agents in 1970 has been estimated at 280 000 t.

Typical *non-ionic surface-active agents* are the condensates formed by ethylene oxide with octylphenol or octylcresol. They are

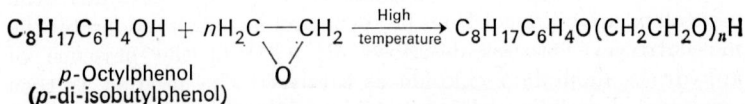

$$C_8H_{17}C_6H_4OH + nH_2C\!\!-\!\!CH_2 \xrightarrow[\text{temperature}]{\text{High}} C_8H_{17}C_6H_4O(CH_2CH_2O)_nH$$

p-Octylphenol
(*p*-di-isobutylphenol)

widely used in households, *e.g.* as washing up liquids, and in the textile industry. World production of non-ionic surface-active agents in 1970 has been estimated as 915 000 t.

Sales of synthetic detergents in the UK in 1969 amounted to 484 800 t. World production of synthetic detergents in 1969 amounted to 7.4 Mt, whereas in 1960 production was only 3.1 Mt. In contrast, world soap production (6.8 Mt in 1969) has not increased much since 1960. Detergent powders contain a number of additives such as sodium tripolyphosphate, sodium sulphate and carboxymethylcellulose. In 1968 proteolytic enzyme-containing powders were introduced which have further increased the sales of synthetic detergents at the expense of soaps. Such detergents are expected to account for about 35 per cent of the UK consumption of detergent powders in 1970. The enzymes are bacterial fermentation products and catalyse the hydrolytic decomposition of proteins (blood and egg stains, *etc.*) into simpler products.

References

1. 'Polypropylene', *Ind. Engng Chem.*, *ind.(int.) Edn*, 1963, **55**(2), 30.
2. G. Natta, *The polymerization and production of olefins*, 2; and *The polymers of symmetrical olefins*, 11, CIBA Review 1964/1965.
3. A. V. Topchiev and B. A. Krentsel, *'Polyolefins'*. Translated from the Russian by A. D. Norris. Oxford: Pergamon 1962.
4. J. C. Swallow, 'Ethylene polymers, past and future', *Trans. J. Plast. Inst.*, 1963, **31**(2), 1.
5. 'Polypropylene after one half a decade', *Chem. Engng Progress*, 1965, **61**(8), 88.
6. H. P. Frank, *Polypropylene*. London: Macdonald Technical and Scientific, 1969.
7. K. Fletcher, R. N. Haward and J. Mann, 'Rubber-reinforced polystyrene and co-polymers', *Chemy Ind.*, 1965, 1854.
8. 'Expanded polystyrene', *Br. Plast.*, 1965, **38**(2), 62.
9. W. S. Penn, *PVC technology*. London: Maclaren, 1966.
10. F. Chevassus and R. de Brontelles, *The stabilization of polyvinyl chloride*. London: Edward Arnold, 1963.

11. M. B. Horn, *Acrylic resins*. New York: Reinhold, 1966.
12. A. Barron and C. R. Patrick, 'Fluorine-containing organic polymers', *Br. Plast.*, 1963, **36**(2), 7.
13. N. J. L. Megson, *Phenolic resin chemistry*. London: Butterworths, 1958.
14. C. P. Vale and W. G. K. Taylor, *Amino plastics*. London: Butterworths, 1964.
15. 'Aminoresins and plastics', *Encyclopedia of chemical technology*, 3rd edn, (P. E. Kirk and D. F. Othmer eds), vol 2, p 225. New York: Interscience, 1963.
16. (a) H. Lee and K. Neville, *Handbook of epoxy resins*. New York: McGraw-Hill, 1966.
    (b) W. G. Potter, *Epoxide resins*. A Plastics Institute Monograph. London: Butterworths, 1970.
17. E. Kilner and D. M. Samuel, *Applied organic chemistry*, p 385. London: Macdonald & Evans, 1960.
18. P. Morgan (ed.), *Glass-reinforced plastics*, 3rd edn. London: Iliffe, 1961.
19. I. A. Eldib, 'How to produce polyurethanes', *Hydrocarb. Process.*, 1963, **42**(12), 121.
20. L. N. Phillips and W. B. V. Parker, *Polyurethanes, Chemical technology and properties*. London: Iliffe, 1964.
21. R. W. Moncrieff, *Man-made fibres*, 4th edn. London: Maclaren, 1966.
22. 'Polyester fibres'—process survey, *Chem. Process Engng*, 1970, **51**(7), 55.
23. A. S. Craig, *Rubber technology, a basic course*. London: Oliver and Boyd, 1963.
24. W. E. Hamer, J. D. Pratt and J. E. Such, *Chemicals for the rubber, leather and matchmaking industries*. Oxford: Pergamon, 1967.
25. M. Morton (ed.), *Introductory rubber technology*. New York: Reinhold, 1959.
26. S. H. Morrell, 'Rubbers today', *Chemy Ind.*, 1962, 106.
27. E. W. Madge, 'The new elastomers', *Chemy Ind.*, 1806.
28. Ref. 17, p 178.
29. R. A. Duckworth, 'Synthetic detergent powders'—process survey, *Chem. Process. Engng*, 1970, **51**(4), 63.

3. Hydrocarbons

Paraffins

Methane for use as a chemical feedstock* is available in the UK in the form of refinery gases (458 000 t produced in 1969) and as North Sea natural gas.[1] Its most important use in the chemical industry is for the production of synthesis gas by steam reforming.[2]

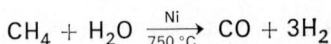

$$CH_4 + H_2O \xrightarrow[750\,°C]{Ni} CO + 3H_2$$

North Sea natural gas is now replacing petroleum naphtha in the UK as the feedstock for the production of synthesis gas (carbon monoxide and hydrogen). This is used for the production of alcohols by hydroformylation (p 62), for methanol synthesis (p 56) and for the production of pure hydrogen for ammonia synthesis. Methane is also used in the UK for the production of hydrogen cyanide, by reaction with ammonia and air over a platinum–rhodium catalyst at 1000 °C (Andrussow process),[3]

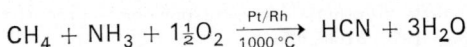

$$CH_4 + NH_3 + 1\tfrac{1}{2}O_2 \xrightarrow[1000\,°C]{Pt/Rh} HCN + 3H_2O$$

and for the production of carbon disulphide by reaction with sulphur in the vapour phase in the presence of a catalyst.[4]

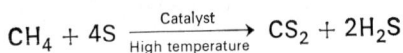

$$CH_4 + 4S \xrightarrow[\text{High temperature}]{\text{Catalyst}} CS_2 + 2H_2S$$

The former process is less important now that hydrogen cyanide is available as a by-product from the production of acrylonitrile (p 87). Both processes are carried out on a relatively small scale.

In the US and in Germany methane is also used for the production of methyl chloride and methylene chloride by thermal chlorination (p 52).

Ethane is important as a source of ethylene in the US. In the UK it is largely used as a fuel but also for the production of ethyl chloride (p 53).

---

* Methane as a component of town gas is dealt with in the companion Industrial Inorganic Chemistry Monograph.

Propane and butanes (LPG) are used in very large quantities as bulk fuels in industry because of their versatility and freedom from sulphur. Important outlets in industry include metal cutting, heat treating, and for generating furnace atmospheres in such processes as gas carbonizing and carbonitriding. They are also used for enriching town gas. Production of propane and butanes in the UK in 1969 amounted to 44 000 t and 777 000 t, respectively. Propane is important as a source of ethylene and propylene (p 6) in the US. Butanes are used for the production of butadiene (p 40). n-Butane is isomerized in petroleum refineries because iso-paraffins have higher octane numbers than n-paraffins.

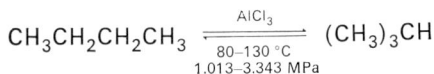

$$CH_3CH_2CH_2CH_3 \xrightarrow[\substack{80-130\ °C \\ 1.013-3.343\ MPa}]{AlCl_3} (CH_3)_3CH$$

The aluminium chloride catalyst is used in the form of a liquid complex with the hydrocarbon and dry hydrogen chloride (promoter); the conversion per pass is about 40 per cent.

Higher n-paraffins in the C_{10}–C_{18} range have become important as raw materials for the production of 'biodegradable' detergents (p 32) and plasticizers (p 20). They are separated from iso-paraffins and aromatic hydrocarbons by the use of synthetic zeolites which act as molecular sieves. The latter will allow n-paraffins with cross-sectional diameters of about 4.9 Å to pass through, but not iso-paraffins, aromatic hydrocarbons and alicyclic hydrocarbons which have cross-sectional diameters of at least 5.6 Å.[5] Another commercial process for isolating n-paraffins makes use of the fact that long-chain n-paraffins, but not branched-chain paraffins and cyclic hydrocarbons, form inclusion compounds with urea. Aliphatic sodium sulphonates with 12–16 carbon atoms in the alkyl groups are manufactured by sulphochlorinating n-paraffins and hydrolysing

$$RH + SO_2 + Cl_2 \xrightarrow{uv\ light} RSO_2Cl + HCl$$

the sulphonyl chlorides with alkali solution.

Higher paraffins were at one time obtained by the Fischer–Tropsch process (the conversion of a mixture of carbon monoxide and hydrogen into hydrocarbons):

$$(2n + 1)H_2 + nCO \longrightarrow C_nH_{2n+2} + nH_2O$$

The only plant now being operated is in South Africa, which has no crude oil reserves but has an ample supply of cheap coal. The process (Sasol process) is a modification of the Fischer–Tropsch process and has been in use for about ten years, producing liquid fuels and organic chemicals. Low-grade bituminous coal is gasified with steam and oxygen under pressure to give synthesis gas, which is freed from hydrogen sulphide and carbon dioxide by treatment under pressure with methanol, and is then split into two streams. One stream is fed to a synthesis unit supplied with a fixed-bed catalyst (precipitated iron) and the other stream is treated with steam and oxygen to convert the hydrocarbons present to carbon monoxide and hydrogen. The resulting mixture, after purification, is fed to a synthesis unit operating with a fluidized catalyst (ferric oxide). The fixed-bed unit produces diesel fuels and waxes, whilst the fluid-bed unit gives motor spirit and oxygenated organic compounds, chiefly lower alcohols and ketones.[6]

Olefins

Ethylene[7]

Production of ethylene in the UK in 1970 amounted to 981 900 t, an increase of 12 per cent over the 1969 production.

Uses of ethylene

Polyethylenes (p 13).   Production of polyethylene accounts for more than half the ethylene consumption.   Production of low density and high density polyethylenes in the UK in 1969 was 240 000 t and 54 000 t, respectively. The polyethylenes and polypropylene accounted for 27 per cent of the total consumption of plastics materials in 1970.   Co-polymers of ethylene and vinyl acetate are tougher and more flexible than low density polyethylene and are now made in the UK.   They are competitors of plasticized polyvinyl chloride over which they have several advantages

Ethylene dihalides and vinyl chloride (p 54).   Production of ethylene dichloride in Western Europe in 1969 was 0.3 Mt

Ethylene oxide is the third biggest outlet for ethylene in the UK.   Production in Western Europe in 1969 was 0.5 Mt

Ethanol (p 57)

Ethylbenzene (for styrene, p 106). Production of ethylbenzene and ethanol in Western Europe in 1969 was 0.3 and 0.29 Mt, respectively

Ethyl chloride (p 53)

Other products. These include acetaldehyde (p 72) and ethylene–propylene terpolymer rubbers (p 32).   Acetaldehyde production from ethylene in Western Europe in 1969 was 0.17 Mt

Propylene[8]

The UK production of propylene in 1970 was 448 060 t, a decrease of 6 per cent on the 1969 production.

Uses of propylene

Production of motor spirit blending components by polymerization (p 32)

Isopropanol (p 59). In 1966, 1.37 billion lb of propylene were used for the manufacture of isopropanol in the US

Acrylonitrile (p 87). The production of acrylonitrile in the US is expected to increase by 250 per cent by 1975

Phenol and acetone (p 121). Phenol, isopropanol and acrylonitrile are probably the substances produced in greatest quantity from propylene

Polypropylene (p 16). Production of polypropylene in the UK in 1967 was 63 000 t. It is important as a plastic material and for the production of fibres. The demand for polypropylene over the next few years is expected to increase considerably

Propylene oxide (p 67)

Butanols (p 62)

Allyl chloride (p 55)

Acrylic acid. The production of acrylic acid by the catalytic air oxidation of propylene is a new use for propylene

Butenes[9]
Uses of n-butenes

Butadiene (p 40). Production of butadiene is by far the most important use for the n-butenes

sec-Butanol (p 62)

Maleic anhydride (p 82). Maleic anhydride is not manufactured from n-butenes in the UK

Uses of isobutene

Butyl rubber (p 29). The production of butyl rubber probably consumes more isobutene than any other use

Polyisobutenes. Polymerization of isobutenes is carried out at low temperatures in the presence of a cationic catalyst, such as anhydrous aluminium chloride or boron trifluoride. The polymer is used as a constituent of viscostatic lubricating oils and in adhesive compositions

Di-isobutenes. These are made by dimerizing isobutene

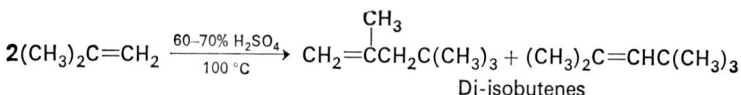

$$2(CH_3)_2C{=}CH_2 \xrightarrow[100\,°C]{60{-}70\%\ H_2SO_4} CH_2{=}\overset{\overset{\displaystyle CH_3}{|}}{C}CH_2C(CH_3)_3 + (CH_3)_2C{=}CHC(CH_3)_3$$

Di-isobutenes

The di-isobutenes are used for the production of 3,5,5-trimethylhexanol (by the hydroformylation process, p 62), for alkylating phenol and cresols to produce octylphenol and octylcresol respectively, and for making anti-oxidants

Other uses. These include the production of t-butanol and p-t-butylphenol

Butadiene

Butadiene is produced in greater quantity than any other organic chemical with the exception of sucrose—world production is now approaching 3 Mt/a and by 1980 it is expected to rise to 4–5 Mt/a. UK production in 1970 was 170 160 t. It has already been mentioned that the major source of ethylene and propylene in Europe is the steam-cracking of naphtha. The amount of butadiene which is also formed depends on the composition of the naphtha and on the cracking temperature. Under severe cracking conditions about 5 per cent of the product consists of butadiene. Naphtha is also the major source of butadiene in Europe but some is also produced by the dehydrogenation of butanes and butenes (the major source of butadiene in the US). Butadiene is recovered from the C_4-fraction by countercurrent extraction with a selective solvent or by extractive distillation. One of the earliest selective solvents (and still in use) was ammoniacal cuprous acetate solution at 5 °C; the butadiene is recovered by heating the solution. Extractive distillation processes are operated by introducing the C_4-feed into the middle of a fractionating column and the solvent into the head of the column. The butenes distil overhead, whilst the butadiene is washed out of the vapour by the solvent and is taken off as base fraction. The first solvent to be used was furfural, but acetonitrile, N-methylpyrrolidone and dimethylformamide are also now used. The various processes in use for the recovery of butadiene from the C_4-fraction have been reviewed.[10]

Butadiene is produced from the n-butenes by vapour-phase dehydrogenation over a catalyst (several combinations are in use; e.g. iron oxide promoted by potassium carbonate and chromium oxide; calcium and magnesium phosphates; chromium oxide–iron oxide–potassium oxide) at 595–675 °C.

$$CH_3CH{=}CHCH_3 \rightleftharpoons CH_2{=}CHCH{=}CH_2 + H_2$$

The feed is diluted with a large excess of steam which acts as a source of heat and, by dilution, reduces the partial pressure of the butenes, the dehydrogenation reaction being favoured at low pressures.[11]

Uses of butadiene[12,13]

SB-R (Styrene–butadiene rubber) production accounts for by far the greatest quantity of butadiene (p 27). Hard SB-R is produced in far smaller quantities

Butadiene–acrylonitrile co-polymer rubbers (p 30)

Polybutadiene rubbers (p 31). These have been manufactured in the UK since 1964

Other uses. Butadiene has become important recently as a starting material for the production of neoprene (p 49). In the US butadiene is believed to be used by one firm for the production of the nylon 66 intermediate, adiponitrile. A recent use is for the production of nylon 12

A minor use is the production of sulpholane,

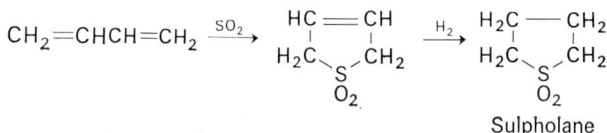

$$CH_2{=}CHCH{=}CH_2 \xrightarrow{SO_2} \begin{array}{c} HC{=\!=\!=}CH \\ | \quad\quad | \\ H_2C\diagdown_{\underset{O_2}{S}}\diagup CH_2 \end{array} \xrightarrow{H_2} \begin{array}{c} H_2C{-\!\!-\!\!-}CH_2 \\ | \quad\quad | \\ H_2C\diagdown_{\underset{O_2}{S}}\diagup CH_2 \end{array}$$

Sulpholane

which is an important selective solvent for aromatic hydrocarbons (see p 101)

ABS polymers (p 88)

Isoprene[14,15]

Isoprene, $CH_2{=}C(CH_3)CH{=}CH_2$, is relatively expensive to produce. It is obtained by the catalytic dehydrogenation of pentanes and/or pentenes which are available in limited quantities in the C_5–fraction from cracking processes, particularly from the steam-cracking of naphtha. It is also manufactured in the US from propylene.

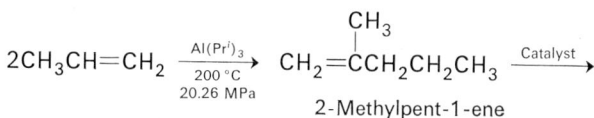

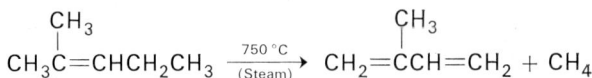

$$2CH_3CH{=}CH_2 \xrightarrow[\substack{200\,°C \\ 20.26\ MPa}]{Al(Pr^i)_3} \begin{array}{c} CH_3 \\ | \\ CH_2{=}CCH_2CH_2CH_3 \end{array} \xrightarrow{Catalyst}$$

2-Methylpent-1-ene

$$\begin{array}{c} CH_3 \\ | \\ CH_3C{=}CHCH_2CH_3 \end{array} \xrightarrow[\text{(Steam)}]{750\,°C} \begin{array}{c} CH_3 \\ | \\ CH_2{=}CCH{=}CH_2 \end{array} + CH_4$$

In a recently developed process isoprene is produced by pyrolysis of the product from the reaction between isobutene and formaldehyde.

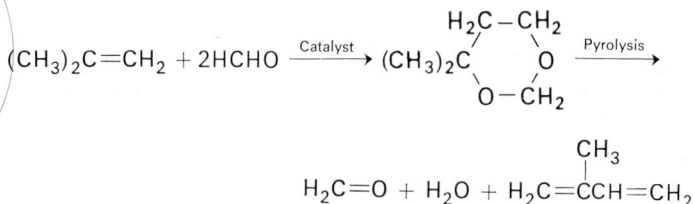

$$(CH_3)_2C{=}CH_2 + 2HCHO \xrightarrow{Catalyst} (CH_3)_2C\underset{\diagdown O - CH_2}{\overset{\diagup H_2C - CH_2}{\diagdown\ \ \ \ O\diagup}} \xrightarrow{Pyrolysis}$$

$$H_2C{=}O + H_2O + \begin{array}{c} CH_3 \\ | \\ H_2C{=}CCH{=}CH_2 \end{array}$$

Isoprene is used as a co-monomer in the production of butyl rubber (p 27) and in the US for the production of high *cis*-polyisoprene rubbers (p 31). Isoprene consumption in the US during 1970 has been estimated as 450 Mlb.

Acetylene

The production of calcium carbide in the UK (168 300 t in 1968; 291 500 t in 1964) is expensive because of the high cost of electricity: power consumption in producing one ton of acetylene by the carbide process is 9000 kW h. Thus, although acetylene has in the past been produced exclusively from calcium carbide in the UK, it will eventually all be produced from hydrocarbons.[16] In 1965 hydrocarbon raw materials, either natural gas or petroleum naphtha, accounted for 50 per cent of the production of acetylene. One of the most important processes, the BASF process (Badische Anilin and Soda Fabrik AG), consists of subjection of natural gas to partial combustion with oxygen (using a specially designed burner), the

$$CH_4 + 1\tfrac{1}{2}O_2 \longrightarrow CO + 2H_2O$$

heat liberated being used to raise the temperature of a further quantity of gas to about 1500 °C (since the decomposition of methane is an endothermic reaction it will be favoured by high temperatures):

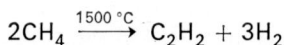

$$2CH_4 \xrightarrow{\;1500\,°C\;} C_2H_2 + 3H_2$$

The reaction time at 1500 °C is 0.001–0.01 s. The products of the process are rapidly cooled (to minimize decomposition of acetylene), compressed and scrubbed with a solvent selective for acetylene.

Figure 8 gives the flow diagram for the BASF process. Oxygen and natural gas (in a mol. ratio of 0.6 : 1.00) are preheated in direct-fired heaters to 649 °C. The burner comprises a mixing section and flame and quench chambers. The mixed gases are fed to the flame room through multiple ports in a burner block designed to prevent backtravel. About one third of the methane is cracked, and most of the remainder is burned. Methane conversion is 90–95 per cent. The acetylene is cooled rapidly by a series of sprays in the lower part of the burner. The cooled gases (containing about 8.5 per cent acetylene) pass to a spray chamber, where most of the water is condensed, and leave the chamber at 38 °C. Soot and polymer-forming compounds are then removed and the cleaned gas is compressed to about 1.013 MPa and scrubbed with a solvent having a high acetylene : carbon dioxide selectivity. Carbon dioxide and acetylene are separated from the solvent in strippers, the acetylene being of 99 per cent purity. The overall yield of acetylene based on carbon in the natural gas is 31 per cent.

FIG. 8. (Facing page) Acetylene (BASF)—Chemical Construction Corp. (Reproduced from the November 1969 issue of *Hydrocarbon Processing*, p 142, by permission of the Gulf Publishing Co., Houston, Texas.)

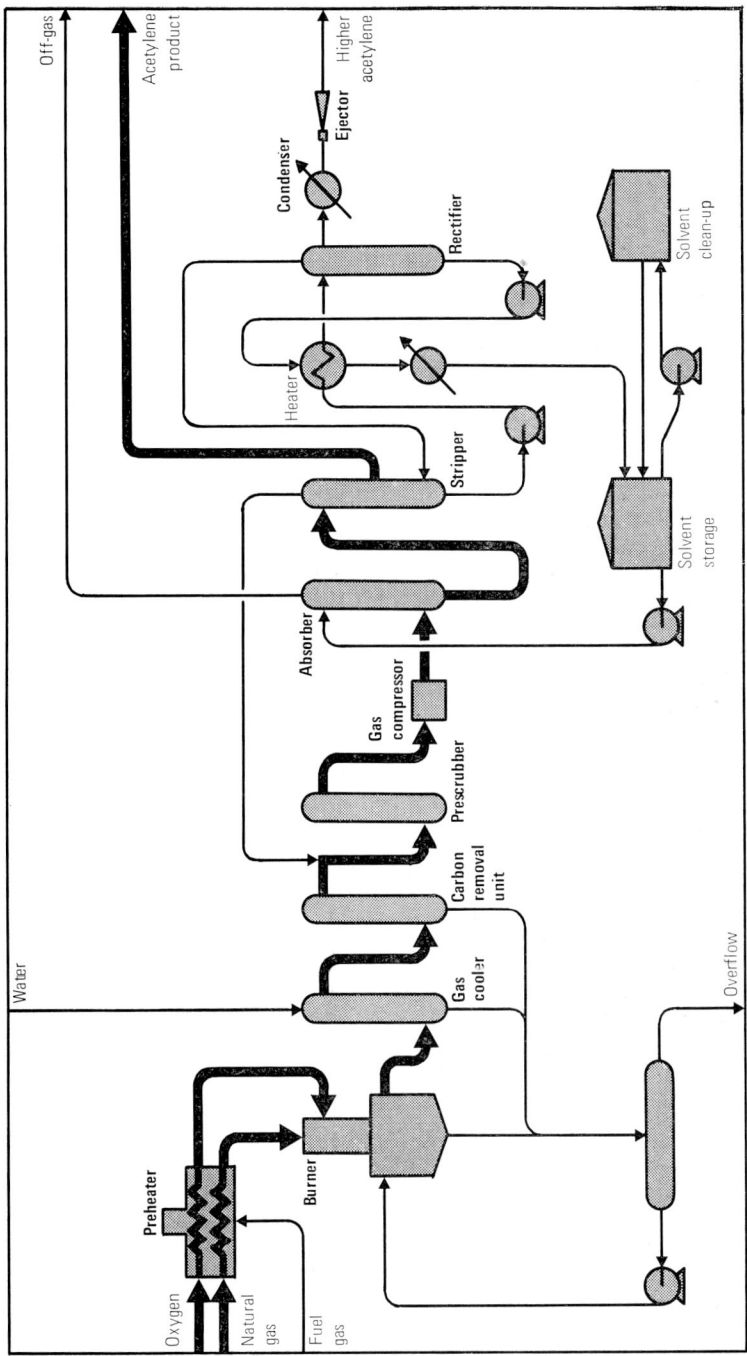

Off-gas

Acetylene product

Higher acetylene

Ejector

Condenser

Rectifier

Heater

Stripper

Absorber

Gas compressor

Prescrubber

Carbon removal unit

Gas cooler

Solvent clean-up

Solvent storage

Water

Burner

Preheater

Oxygen

Natural gas

Fuel gas

Overflow

In the Wulff process hydrocarbons are cracked in a refractory-lined furnace previously heated by burning fuel to 900–1300 °C. Two furnaces are operated cyclically, cracking taking place in one whilst the other is being heated. The acetylene is diluted with steam and the cracking time is less than 0.05 s. As in the partial oxidation process the acetylene is recovered by extraction with a selective solvent, e.g. dimethylformamide. Acetylene is being produced by both the Wulff and BASF processes in the UK, using naphtha as feedstock.

Acetylene has lost much of its former importance as a raw material for the production of other organic compounds owing to competition from the olefins. Ethylene and propylene are produced more cheaply than acetylene from carbide or hydrocarbons (the price of ethylene is about half that of acetylene), since the steam-cracking of naphtha lends itself to very large scale production. However, processes which yield a high ratio of ethylene to acetylene can be operated on a larger scale and are more competitive with ethylene production by naphtha cracking. Ethylene has another advantage over acetylene in that it can readily be pumped through pipelines under pressure (e.g. there is a pipeline from the Esso Refinery at Fawley, Southampton, to the ICI Severnside works near Bristol).[17]

Products based on tetrachloroethane

Some organic chemicals have long been produced almost exclusively from acetylene, such as products based on tetrachloroethane,[18,19] alkyl vinyl ethers, CH_2=CHOR, vinylpyrrolidone and vinylcarbazole. The most important of these substances are those products based on tetrachloroethane. The latter is produced by passing acetylene and chlorine into tetrachloroethane containing a suspension of anhydrous ferric chloride (catalyst) at about 70 °C:

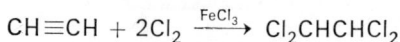

$$CH \equiv CH + 2Cl_2 \xrightarrow{FeCl_3} Cl_2CHCHCl_2$$

Under these conditions the explosive reaction of acetylene with chlorine to give carbon and hydrogen chloride is suppressed. Tetrachloroethane itself is used solely as a starting compound for the production of the following compounds (it is one of the most toxic of the chlorinated solvents):

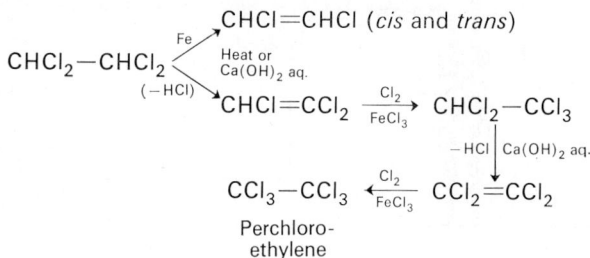

$$CHCl_2-CHCl_2 \xrightarrow[(-HCl)]{Fe}$$
$$CHCl=CHCl \ (\textit{cis and trans})$$
$$\xrightarrow[\text{Ca(OH)}_2 \text{ aq.}]{\text{Heat or}}$$
$$CHCl=CCl_2 \xrightarrow[FeCl_3]{Cl_2} CHCl_2-CCl_3$$
$$-HCl \downarrow Ca(OH)_2 \text{ aq.}$$
$$CCl_3-CCl_3 \xleftarrow[FeCl_3]{Cl_2} CCl_2=CCl_2$$

Perchloro-
ethylene

Perchloroethylene is also manufactured by the pyrolysis of carbon tetrachloride (at 800–900 °C) or pentachloroethane:

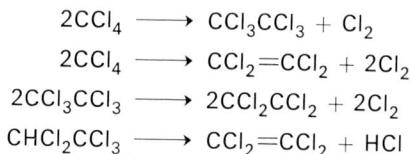

$$2CCl_4 \longrightarrow CCl_3CCl_3 + Cl_2$$
$$2CCl_4 \longrightarrow CCl_2{=}CCl_2 + 2Cl_2$$
$$2CCl_3CCl_3 \longrightarrow 2CCl_2CCl_2 + 2Cl_2$$
$$CHCl_2CCl_3 \longrightarrow CCl_2{=}CCl_2 + HCl$$

The latter process is carried out at 250 °C over a catalyst of barium chloride on activated charcoal; the hydrogen chloride produced can be used for the production of vinyl chloride or ethyl chloride. Perchloroethylene now accounts for about 85 per cent of the solvents used in the dry cleaning industry in the UK; it readily dissolves oily soil from clothes and has no effect on textile fibres. It is also non-flammable and suitable for use in automatically controlled machines.

Trichloroethylene is usually made now by the dehydrochlorination of tetrachloroethane at 500 °C, a process which is superceding treatment of the latter compound with lime:

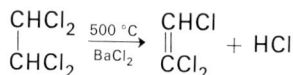

$$\begin{array}{c} CHCl_2 \\ | \\ CHCl_2 \end{array} \xrightarrow[BaCl_2]{500\ °C} \begin{array}{c} CHCl \\ \| \\ CCl_2 \end{array} + HCl$$

It is most widely used as a metal degreasing solvent but also as a solvent for oils and fats, as a component of paint strippers and as a general anaesthetic (the anaesthesia is not deep) for short operations. It is currently used only to a small extent in dry cleaning because of its effect (loss of colour) on cellulose triacetate fibre. Both tetrachloroethylene and trichloroethylene have toxic properties, causing gradual deterioration of internal organs on continued exposure to low concentrations.

Vinyl chloride[20,21]
Vinyl chloride, $CH_2{=}CHCl$, is still produced by the pre-World War II process of addition of hydrogen chloride to acetylene.

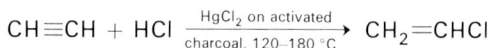

$$CH{\equiv}CH + HCl \xrightarrow[\text{charcoal, 120–180 °C}]{\text{HgCl}_2 \text{ on activated}} CH_2{=}CHCl$$

The chief disadvantage of this process is the cost of the acetylene; in other respects it is satisfactory. Anhydrous hydrogen chloride must be available on the same site, *e.g.* from the thermal cracking of tetrachloroethane to produce trichloroethylene or from the production of vinyl chloride from ethylene dichloride:

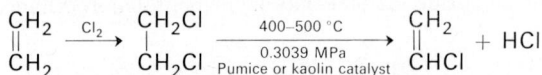

$$\begin{array}{c}CH_2 \\ \| \\ CH_2\end{array} \xrightarrow{Cl_2} \begin{array}{c}CH_2Cl \\ | \\ CH_2Cl\end{array} \xrightarrow[\substack{0.3039\ MPa \\ \text{Pumice or kaolin catalyst}}]{400-500\ ^\circ C} \begin{array}{c}CH_2 \\ \| \\ CHCl\end{array} + HCl$$

In order to make the latter process economical the hydrogen chloride produced must be profitably disposed of either by sale, by using it in another process (as above) or for making ethylene dichloride (*oxyhydrochlorination*).

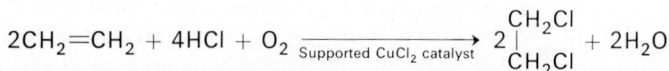

$$2CH_2{=}CH_2 + 4HCl + O_2 \xrightarrow{\text{Supported CuCl}_2\text{ catalyst}} 2\begin{array}{c}CH_2Cl \\ | \\ CH_2Cl\end{array} + 2H_2O$$

The use of the hydrogen chloride for the above purpose is the ideal solution and since 1965 vinyl chloride plants have been based on ethylene with the incorporation of oxyhydrochlorination. In 1970 about 80 per cent of vinyl chloride production in Western Europe was based on ethylene.

Figure 9 is a flow diagram for the production of ethylene dichloride (EDC) by addition and oxychlorination and of vinyl chloride from EDC by cracking. The addition reaction takes place in a liquid-phase reactor, whilst oxychlorination is carried out in the vapour phase in the presence of an efficient catalyst. High pressure steam generated in the oxychlorination reactor removes the heat of reaction. The crude EDC produced by both processes is combined with recycle EDC from the cracking unit and fractionated to remove light and heavy ends. The effluent from the cracking furnace is quenched, separated into hydrogen chloride (which is recycled to the oxychlorination unit), vinyl chloride and unchanged EDC (which is recycled).

Vinyl chloride is largely used for the production of polyvinyl chloride (p 19) but is also co-polymerized with several other monomers. Production of polyvinyl chloride in the UK in 1969 was 258 900 t.

Vinyl acetate
Vinyl acetate, $CH_2{=}CHOCOCH_3$, is at present made from acetylene[22] and ethylene[23] and, as yet, the acetylene process appears to be holding its own. The latter is carried out in the vapour phase:

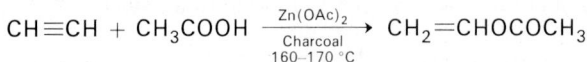

$$CH{\equiv}CH + CH_3COOH \xrightarrow[\substack{\text{Charcoal} \\ 160-170\ ^\circ C}]{Zn(OAc)_2} CH_2{=}CHOCOCH_3$$

The ethylene-based process consists of passing ethylene and oxygen into an acetic acid solution containing palladium chloride, an

---

Fig. 9. (Facing page) Vinyl chloride—Stauffer Chemical Co. (Reproduced from the November 1969 issue of *Hydrocarbon Processing*, p 249, by permission of the Gulf Publishing Co., Houston, Texas.)

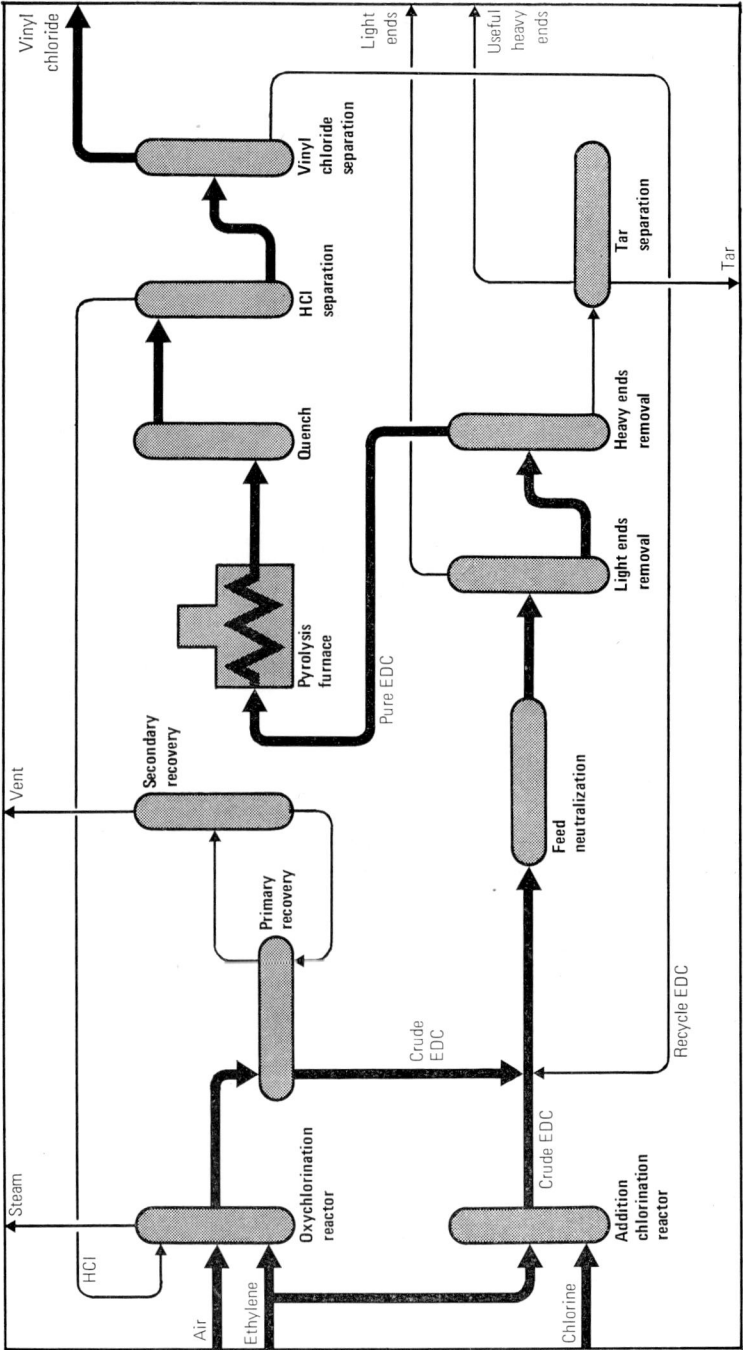

Air

Ethylene

Steam

HCl

Oxychlorination reactor

Chlorine

Addition chlorination reactor

Crude EDC

Crude EDC

Recycle EDC

Primary recovery

Secondary recovery

Vent

Feed neutralization

Light ends removal

Pyrolysis furnace

Pure EDC

Quench

Heavy ends removal

Tar separation

Tar

HCl separation

Vinyl chloride separation

Light ends

Useful heavy ends

Vinyl chloride

FIG. 10. A Horton's sphere used for storing vinyl chloride prior to polymerization. (By courtesy of BP Chemicals International Ltd.)

alkali metal acetate and cupric ions. The palladium is oxidized

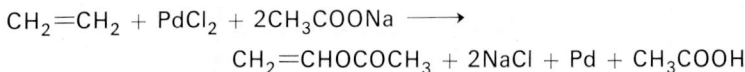

$$CH_2{=}CH_2 + PdCl_2 + 2CH_3COONa \longrightarrow$$
$$CH_2{=}CHOCOCH_3 + 2NaCl + Pd + CH_3COOH$$

by the cupric ions back to palladium chloride, thus making the process continuous (*see also* manufacture of acetaldehyde from ethylene, p 72). The chief attraction of the process is the lower cost of the ethylene feedstock compared with acetylene. However, both the capital cost and the energy consumption of vinyl acetate plants based on ethylene is higher than those based on acetylene. World production of vinyl acetate is now probably about 1 Mt/a.

Polyvinyl acetate is important in the UK as the essential constituent in plastic emulsion paints; it is also used as an adhesive and in the textile industry. Production of polyvinyl acetate in the UK in 1969 was 38 700 t, an increase of 4300 t over the 1968 figure. Vinyl acetate is also used for the manufacture of polyvinyl butyral and polyvinyl alcohol and is co-polymerized with vinyl chloride. Polyvinyl alcohol is obtained by subjecting polyvinyl acetate to methanolysis (see transesterification p 84) in the presence of an alkaline or acid catalyst.

Acrylonitrile
Acrylonitrile, CH_2=CHCN, was formerly made largely by the catalytic addition of hydrogen cyanide to acetylene, and also from ethylene oxide. It is now made much more cheaply by the ammo-oxidation of propylene, a process which is also safer and easier to operate (p 87).

Chloroprene
Chloroprene, CH_2=CHC(Cl)=CH_2, is the monomer used for the production of the important synthetic rubber, neoprene (p 31). The

$$2CH\equiv CH \xrightarrow[55-70\,°C]{Cu_2Cl_2/NH_4Cl} CH_2=CHC\equiv CH$$

$$\xrightarrow[Cu_2Cl_2/NH_4Cl]{+HCl} CH_2=CHC(Cl)=CH_2$$
Chloroprene

latter has been an acetylene-based rubber since its discovery in 1933. However, in recent years a new process has been discovered (BP Chemicals International Ltd) based on butadiene which will probably become the preferred route.

$$CH_2=CHCH=CH_2 \xrightarrow[400\,°C]{Cl_2} CH_2=CHCHClCH_2Cl \quad bp\ 123\,°C$$

$$ClCH_2CH=CHCH_2Cl \quad cis\ \ bp\ 154\,°C$$
$$trans\ bp\ 157\,°C$$

(Catalytic isomerization of 1,4-dichlorobutene to 3,4-dichlorobutene)

Cu^{2+}

$$CH_2=CHC(Cl)=CH_2 \xleftarrow[-HCl]{NaOH\ aq.} CH_2=CHCHClCH_2Cl$$

Separated from the catalysed mixture by fractional distillation

The greater part of the world production of acrylic acid and acrylates is still based on acetylene, but a new process has recently been brought into production using propylene as raw material (p 82).

The production of acetaldehyde from acetylene is now of little

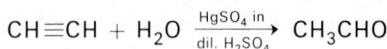

$$CH\equiv CH + H_2O \xrightarrow[\text{dil. }H_2SO_4]{\text{HgSO}_4 \text{ in}} CH_3CHO$$

importance. Acetaldehyde is more cheaply produced from ethanol by catalytic dehydrogenation or by the even more attractive palladium chloride process (p 72).

Mention should also be made of the techniques developed in Germany just before World War II for carrying out reactions of acetylene at elevated temperature and pressure (previously considered dangerous). When the gas can be used at moderate pressures it is only necessary to dilute with an inert gas such as nitrogen or steam (which decreases the partial pressure of acetylene). If higher pressures have to be used it is necessary to cut down the amount of free space in the plant to a minimum (*e.g.* all wide-diameter tubes have to be packed with bundles of narrow-diameter tubes). An example of the use of acetylene under pressure is the production of butane-1,4-diol and tetrahydrofuran.

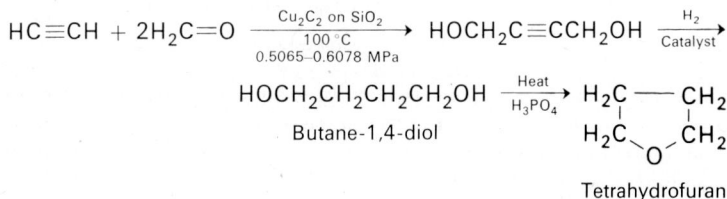

$$HC\equiv CH + 2H_2C=O \xrightarrow[\text{0.5065–0.6078 MPa}]{\substack{\text{Cu}_2\text{C}_2 \text{ on SiO}_2 \\ 100\ ^\circ\text{C}}} HOCH_2C\equiv CCH_2OH \xrightarrow[\text{Catalyst}]{H_2}$$

$$\underset{\text{Butane-1,4-diol}}{HOCH_2CH_2CH_2CH_2OH} \xrightarrow[H_3PO_4]{\text{Heat}} \begin{array}{c} H_2C{-}\!\!-{-}CH_2 \\ |\qquad\quad| \\ H_2C{\diagdown}_O{\diagup}CH_2 \end{array}$$

Tetrahydrofuran

A good account of the production of organic chemicals from acetylene has been published.[24]

References

1. 'North sea natural gas', *Industrial chemistry—inorganic*, Monographs for Teachers, No. 10, 2nd edn, p 18. London: Royal Institute of Chemistry, 1970.
2. Ref. 1, 'Steam reforming of methane', p 22.
3. 'Hydrogen cyanide–Andrussow process–the Girdler Corp.', *Hydrocarb. Process*, 1961, **40**(11), 253.
4. Ref. 1, 'Carbon disulphide', p 44.
5. 'BP normal paraffins plant', *Chem. Trade J.*, 1964, **155**, 665.
6. P. E. Rousseau, 'Organic chemicals and the Fischer–Tropsch synthesis in South Africa', *Chemy Ind.*, 1962, 1958.
7. R. A. Duckworth, 'Ethylene', *Chem. Process. Engng*, 1968, **49**(2), 67.
8. 'The growing importance of propylene', *Chemy Ind.*, 1967, 774.

9. 'Separation and purification of butenes', *Encyclopedia of chemical technology*, 3rd edn (P. E. Kirk and D. F. Othmer eds), vol 3, p 836. New York: Interscience, 1964.

10. T. Reis, 'Butadiene extraction—process survey', *Chem. Process. Engng*, 1970, **51**(3), 65.

11. J. A. R. Bennett, 'The production of butadiene', *Chemy Ind*, 1961, 410.

12. R. B. Stobaugh, 'Butadiene: how, where, who—future', *Hydrocarb. Process.*, 1967, **46**(6), 41.

13. M. Niel, 'The European butadiene market', ECMRA *conference on heavy chemicals and their raw materials in Europe*, pp 65–80. London: ECMRA, 1969.

14. *Chem. Engng News*, 1967, April 3, 60.

15. R. B. Stobaugh, 'Isoprene: how, where, who—future', *Hydrocarb. Process.*, 1967, **46**(7), 149.

16. D. W. F. Hardie, *Acetylene: manufacture and uses.* London: Oxford University Press, 1965; 'Wulff acetylene—process costs', *Chem. Process. Engng*, 1966, **47**(2), 71.

17. 'Acetylene and ethylene processes—Conference report', *Chem. Process. Engng*, 1968, **49**(5), 101.

18. 'Trichloroethylene', ref. 9, vol 5, p 180. New York: Interscience, 1964.

19. 'Perchloroethylene', *ibid*, 195.

20. L. B. Albright, 'Vinyl chloride processes', *Chem. Engng*, 1967, **74**(3), 127.

21. P. G. Caudle, 'Acetylene or ethylene as raw materials for vinyl chloride—a review of economic factors', *Chemy Ind.*, 1968, 1551.

22. O. Horn, 'Monomeric and polymeric vinyl acetate and derivatives', *Chemy Ind.*, 1955, 1748.

23. 'Vinyl acetate *via* ethylene—process costs', *Chem. Process. Engng*, 1967, 48(3), 71.

24. S. A. Miller, 'Chemicals from acetylene', *Chemy Ind.*, 1963, 14.

4. Halogen-containing Compounds

Alkyl halides

Methyl chloride[1,2] is manufactured in the UK by the reaction between methanol and hydrogen chloride in the presence of a catalyst (*e.g.* zinc chloride on pumice), and in other countries also by the chlorination of methane (5–10:1 part chlorine) at about 400 °C.

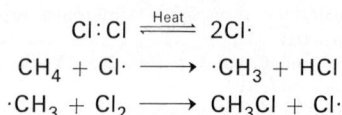

$$Cl:Cl \xrightleftharpoons{\text{Heat}} 2Cl\cdot$$

$$CH_4 + Cl\cdot \longrightarrow \cdot CH_3 + HCl$$

$$\cdot CH_3 + Cl_2 \longrightarrow CH_3Cl + Cl\cdot$$

The formation of higher chlorination products is minimized by using a large excess of methane. Methyl chloride is particularly toxic. Quite low concentrations can cause damage to the central nervous system and internal organs, whilst high concentrations can be fatal.

Uses of methyl chloride

Tetramethyl lead (see p 53)

Methylene chloride and chloroform by vapour phase chlorination in the presence of a catalyst. Methylene chloride is important as a constituent of paint removers and as a diluent in the acetylation of cellulose

Methylation of cellulose to produce methylcelluloses, which are particularly important in textile processes, in cosmetics and for sizing paper

Silicone polymers (fluids, rubbers and resins)

As a catalyst/solvent in the manufacture of butyl rubber

Dimethyl sulphoxide:

$$CH_3Cl \xrightarrow{Na_2S} (CH_3)_2S \xrightarrow[\text{catalytic oxidation}]{\text{Liquid phase}} (CH_3)_2SO_2$$

The latter is used as a selective solvent for aromatic hydrocarbons and as a solvent for dissolving acrylic polymers for spinning into fibres

Methyl bromide is manufactured by the reaction between methanol and hydrogen bromide, and is important as a grain fumigating agent. Its toxic effects are similar to those of methyl chloride.

Ethyl chloride is manufactured in the UK by two integrated processes represented by the following equations:

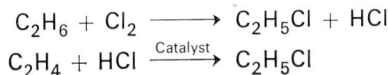

$$C_2H_6 + Cl_2 \longrightarrow C_2H_5Cl + HCl$$
$$C_2H_4 + HCl \xrightarrow{\text{Catalyst}} C_2H_5Cl$$

The anhydrous hydrogen chloride produced in the first process is used in the second process. The chlorination of ethane is carried out at about 400 °C with the ethane in large excess. The process is feasible for obtaining ethyl chloride as chief product since the latter chlorinates at a much slower rate than ethane. It can be carried out on the C_2-stream (ethane and ethylene) from cracking operations: ethylene does not add on chlorine at 400 °C. The addition reaction may be carried out in the liquid phase at 35–40 °C, 3.039–4.052 MPa pressure (to keep the ethyl chloride in the liquid phase) and in the presence of anhydrous aluminium chloride as catalyst with liquid ethyl chloride as solvent, or in the vapour phase over anhydrous aluminium chloride at 130–250 °C.

Annual production of ethyl chloride in the UK has been estimated at about 40 000 t. Ethyl chloride is largely used for making tetraethyl lead, much smaller amounts being used in ethylcellulose manufacture, and as a local and general anaesthetic.

Ethyl bromide is used as an intermediate in the production of barbiturate drugs. The Dow Chemical Co. (US) has developed a process for the manufacture of ethyl bromide which is believed to be the first commercial process in which a radioactive isotope is used for catalysis. Ethylene is treated with hydrogen bromide at 0 °C in the presence of γ-radiation from a cobalt-60 source.[3]

n-Butyl bromide is used for the production of organotin compounds for use as stabilizers for polyvinyl chloride and as fungicides.

Tetraethyl lead

Tetraethyl lead has been manufactured for a long time by heating ethyl chloride with a lead–sodium alloy in an autoclave:

$$4C_2H_5Cl + 4Na/Pb \longrightarrow (C_2H_5)_4Pb + 3Pb + 4NaCl$$

It is also now being manufactured in the US by the electrolysis of an ethereal solution of ethyl magnesium chloride:

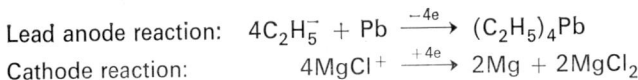

Lead anode reaction: $\quad 4C_2H_5^- + Pb \xrightarrow{-4e} (C_2H_5)_4Pb$

Cathode reaction: $\quad\quad\quad 4MgCl^+ \xrightarrow{+4e} 2Mg + 2MgCl_2$

The process is also used for making tetramethyl lead from methyl magnesium chloride. Tetraethyl and tetramethyl lead are both toxic by inhalation of the vapours and by absorption through the skin.

Ethylene dihalides

Ethylene dichloride is made by the direct addition of chlorine to ethylene and from ethylene by oxyhydrochlorination (p 46). The former process is carried out in liquid ethylene dichloride in the presence of a catalyst (*e.g.* $FeCl_3$). The most important use for ethylene dichloride is the production of vinyl chloride. Other applications include its use as a component of antiknock fluids, as an extracting solvent for oils and fats, for metal degreasing, in rubber cements and in paint strippers, and for the manufacture of ethylene diamine. It has irritant and narcotic properties and can cause damage to internal organs.

Ethylene dibromide is made by passing ethylene into liquid bromine with cooling to remove the heat of reaction and is used mainly as a component of antiknock fluids and to a small extent in agricultural formulations. The toxicity is similar to that of the dichloride.

Chloroform

Chloroform[4] is manufactured by the chlorination of methane or methyl chloride and also by the action of bleaching powder on acetone. In addition to its minor use as a solvent it is important as the raw material for the production of polytetrafluoroethylene (p 19):

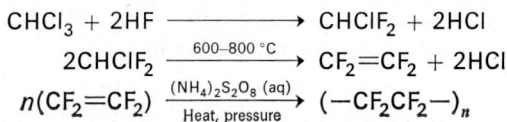

$$CHCl_3 + 2HF \longrightarrow CHClF_2 + 2HCl$$

$$2CHClF_2 \xrightarrow{600-800\ °C} CF_2{=}CF_2 + 2HCl$$

$$n(CF_2{=}CF_2) \xrightarrow[\text{Heat, pressure}]{(NH_4)_2S_2O_8\ (aq)} (-CF_2CF_2-)_n$$

Chloroform can cause damage to internal organs.

Carbon tetrachloride

Carbon tetrachloride[5,6] is manufactured largely from carbon disulphide and chlorine:

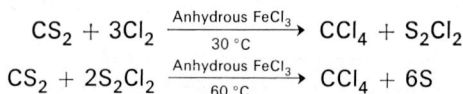

$$CS_2 + 3Cl_2 \xrightarrow[\text{30 °C}]{\text{Anhydrous FeCl}_3} CCl_4 + S_2Cl_2$$

$$CS_2 + 2S_2Cl_2 \xrightarrow[\text{60 °C}]{\text{Anhydrous FeCl}_3} CCl_4 + 6S$$

Alternative, though less important, preparations are the chlorination of methane and the chlorinolysis of propane:

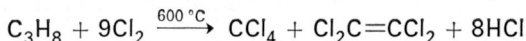

$$C_3H_8 + 9Cl_2 \xrightarrow{600\ °C} CCl_4 + Cl_2C{=}CCl_2 + 8HCl$$

Production of carbon tetrachloride in the US in 1968 was 760 M lb.

It is used largely for the production of *dichlorodifluoromethane*:

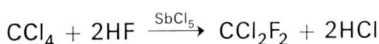

$$CCl_4 + 2HF \xrightarrow{SbCl_5} CCl_2F_2 + 2HCl$$

The actual catalyst is probably $SbCl_2F_3$, which is formed *in situ.*

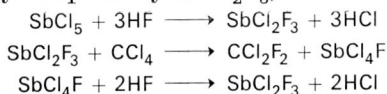

$$SbCl_5 + 3HF \longrightarrow SbCl_2F_3 + 3HCl$$
$$SbCl_2F_3 + CCl_4 \longrightarrow CCl_2F_2 + SbCl_4F$$
$$SbCl_4F + 2HF \longrightarrow SbCl_2F_3 + 2HCl$$

Dichlorodifluoromethane has long been important as a refrigerant, and more recently has become widely used as a propellant for aerosol sprays. Chlorofluoromethane is also manufactured. Carbon tetrachloride is also used as a grain fumigant, in fire extinguishers, as a solvent and for treating liver fluke in sheep and cattle. Because of the toxicity of its vapour it is no longer allowed in household products in the US.

Several other chlorinated hydrocarbons have already been mentioned under acetylene (p 44).

Allyl chloride

Allyl chloride is produced by the chlorination of propylene at about 500 °C (contact time 2 s):

$$CH_3CH{=}CH_2 \xrightarrow[500\ °C]{Cl_2} ClCH_2CH{=}CH_2 + HCl$$

At 500 °C chlorine radicals are formed which attack the methyl groups to give allyl radicals (the allyl radical, being a resonance hybrid of two canonical forms of equal energy content, is the most stable radical which can be formed from propylene).

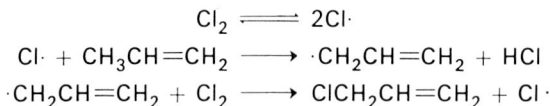

$$Cl_2 \rightleftharpoons 2Cl\cdot$$
$$Cl\cdot + CH_3CH{=}CH_2 \longrightarrow \cdot CH_2CH{=}CH_2 + HCl$$
$$\cdot CH_2CH{=}CH_2 + Cl_2 \longrightarrow ClCH_2CH{=}CH_2 + Cl\cdot$$

Allyl chloride is important as an intermediate for the production of glycerol (p 63) and epichlorohydrin, the latter being used to produce epoxy resins (p 25). It has been manufactured in the UK. Allyl chloride, as well as being an irritant, can damage internal organs.

References

1. E. Kilner and D. M. Samuel, *Applied organic chemistry*, p 45. London: Macdonald and Evans, 1960.
2. W. L. Faith, D. B. Keyes and R. L. Clark, *Industrial chemicals*, 3rd edn, p 507. New York: Wiley, 1965.
3. 'Manufacture of ethyl bromide', *Chem. Trade J.*, 1964, **155**, 705.
4. S. A. Miller, 'Chloroform and carbon tetrachloride', *Chem. Process. Engng*, 1967, **48**(4), 79.
5. Ref. 4.
6. 'Uses of carbon tetrachloride', *Chem. Engng News*, 1963, **41**, 40.

5. Alcohols

Methanol

Methanol was formerly produced almost exclusively by the reaction between hydrogen and carbon monoxide at 25.33–30.39 MPa over a catalyst consisting of zinc oxide (three parts) and chromium oxide (one part) at 300–400 °C. The process is favoured by increased pressure and adversely affected (being exothermic) by increased

$$2H_2 + CO \rightleftharpoons CH_3OH \qquad \Delta H = -94.47 \text{ kJ mol}^{-1}$$

temperature. Thus at 350 °C and 10.13 MPa the percentage methanol at equilibrium is only 3.7 per cent but at 350 °C and 30.39 MPa it is 19.5 per cent. The unchanged gaseous mixture is recycled; part of it is purged to atmosphere from time to time to reduce the accumulation of inert gas (chiefly methane) produced by the reaction $CO + 3H_2 \rightarrow CH_4 + H_2O$. The synthesis gas was formerly obtained from water gas in the UK but is now produced from natural gas by steam-reforming.[1,2]

Recently ICI have developed a new copper catalyst which is sufficiently active to allow the process to take place at a lower temperature (270 °C), counteracting the effect on the equilibrium concentration of methanol of reducing the pressure to 5.065 MPa. At 5.065 MPa the equilibrium concentration of methanol is approximately 2.5 per cent. The main disadvantage of a copper catalyst is that it is easily poisoned by sulphur. This has been overcome, since sulphur-free synthesis gas is obtained from the steam-reforming of naphtha or methane. The lower pressure employed enables centrifugal synthesis gas compressors and circulators with steam turbine drives to be used in all but the smallest plants, with consequent reduction in capital and running costs. The low pressure process also produces purer methanol than the high pressure process (about 2 per cent of the product in the case of the high pressure process consists of dimethyl ether) but, in general, the distillation plant is similar. The economies achieved by the low impurity levels of the crude methanol are: saving of 8–12 per cent in feedstock requirements over the HP process, consequent saving in fuel, and reduction in capital cost for reformer and associated plant.[3,4]

The ICI low pressure methanol synthesis process is illustrated in *Fig. 11*. The feedstock is desulphurized before being passed to a steam reformer, which is designed to give a low residual methane concentration. The heat from the flue and synthesis gas is used to raise high pressure steam for driving the compressor and circulator. The reformed gas is then compressed to 5.065 MPa. If the feedstock is natural gas it is necessary to introduce additional carbon dioxide at the suction of the synthesis gas compressor to maintain the correct C/H ratio in the loop ($CO_2 + 3H_2 \rightleftharpoons CH_3OH + H_2$). The compressed gas passes to the synthesis loop where the basic arrangement is the circulator, converter (of much simpler design than a high pressure converter) with feed-effluent exchanger, condenser and catchpot. Two distillation columns are used to purify the methanol.

Methanol is also produced in the UK at two plants (one of which is modern) by the distillation of hardwood in the absence of air. The process only accounts for a very small proportion of the methanol produced but provides more than one-third of the requirements of wood charcoal in the UK.[5]

Uses of methanol

Formaldehyde (p 70). In the UK 50–60 per cent of the methanol produced is used for making formaldehyde. The demand for methanol for formaldehyde production is increasing

Other compounds. These include esters such as dimethyl terephthalate (for Terylene), methyl methacrylate (for Perspex), methyl halides, methylamines and other methyl derivatives. A new use is for the production of acetic acid

Miscellaneous uses. As a solvent in spirit varnishes, polishes and stains, for denaturing industrial alcohol, in antifreeze compositions, and as a component of motor-spirit blends

The UK is the fourth largest producer of methanol; plant capacity in 1970 was 135 Mgall.

Methanol is toxic when taken orally, by absorption through the skin and by inhalation of the vapour. One of its effects is destruction of the optic nerve (which results in blindness).

Ethanol

Although industrial ethanol is still produced in the UK from molasses by fermentation with yeast, most of the production is from ethylene. Its production from ethylene by absorption of the latter in concentrated sulphuric acid followed by hydrolysis of the products has been operated in the UK since 1942. This process is unattractive because of the large quantity of sulphuric acid involved and the necessity for concentrating the dilute sulphuric acid produced in the course of hydrolysis. Since 1951 ethanol has been produced in the

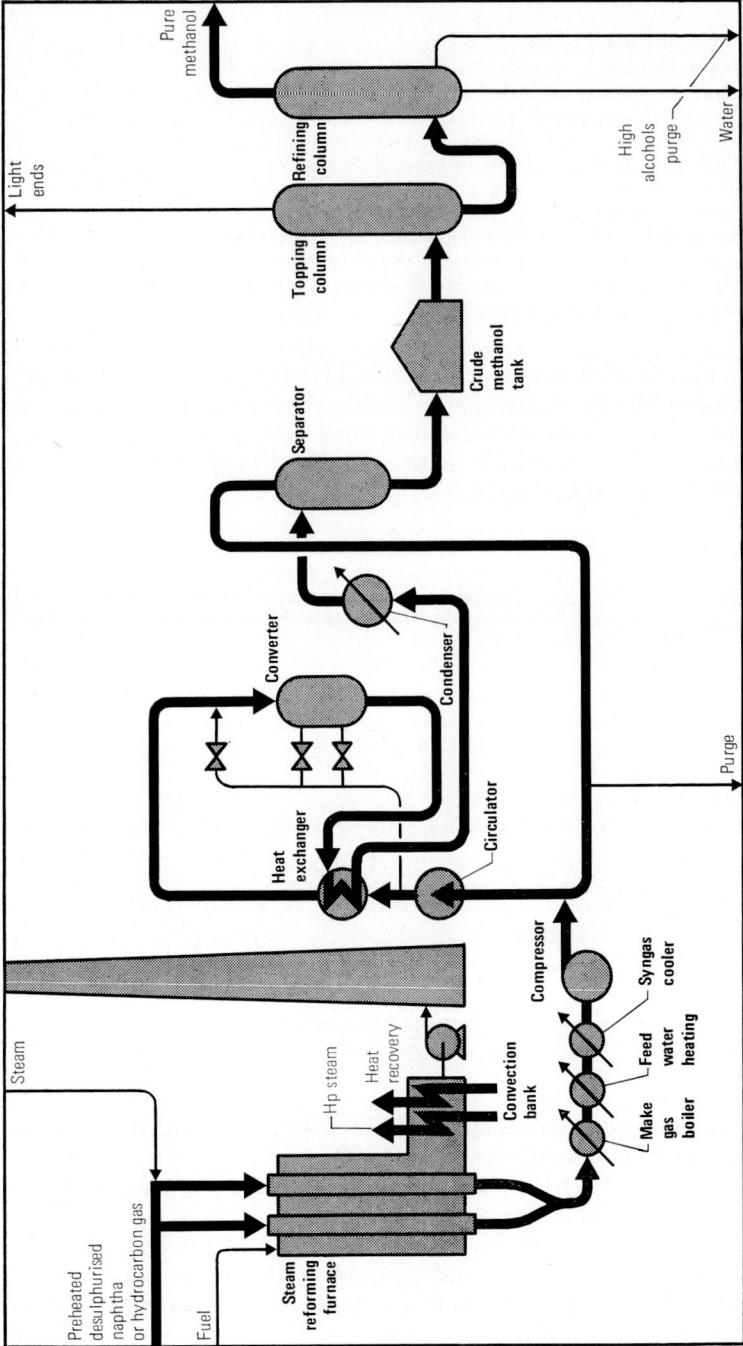

Steam

Preheated desulphurised naphtha or hydrocarbon gas

Fuel

Steam reforming furnace

Hp steam

Heat recovery

Convection bank

Make gas boiler

Feed water heating

Syngas cooler

Compressor

Heat exchanger

Converter

Circulator

Condenser

Purge

Separator

Crude methanol tank

Topping column

Refining column

Light ends

Pure methanol

High alcohols purge

Water

UK by the direct hydration of ethylene.[6,7]

$$CH_2{=}CH_2 + H_2O \longrightarrow CH_3CH_2OH \qquad \varDelta H = -44.10 \text{ kJ mol}^{-1}$$

As in the case of the methanol synthesis the process is favoured by increased pressure (6.787 MPa) and proceeds at a sufficiently fast rate at about 300 °C. It is carried out over a phosphoric acid on Celite (a proprietary brand of kieselguhr) catalyst. The conversion per pass is only about 5 per cent which means that a high proportion of the ethylene feedstock has to be recycled. The reaction mechanism is probably as follows.

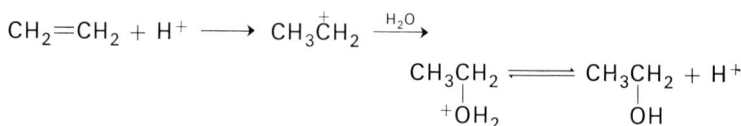

$$CH_2{=}CH_2 + H^+ \longrightarrow CH_3\overset{+}{C}H_2 \xrightarrow{\ H_2O\ }$$

$$CH_3CH_2 \rightleftharpoons CH_3CH_2 + H^+$$
$$\underset{{}^+OH_2}{|} \qquad\qquad \underset{OH}{|}$$

Some diethyl ether is also formed which is recovered. Alternatively,

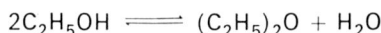

$$2C_2H_5OH \rightleftharpoons (C_2H_5)_2O + H_2O$$

it may be recycled to displace the equilibrium in favour of ethanol. The small amount of acetylene impurity present in the ethylene feedstock gives rise to acetaldehyde, which can, however, be

$$CH{\equiv}CH + H_2O \xrightarrow{\text{Catalyst}} CH_3CHO$$

converted to ethanol by catalytic hydrogenation. The ethanol is condensed as a dilute aqueous solution and concentrated in a fractionating column. The concentrate is hydrogenated over a nickel-on-kieselguhr catalyst to about 85 per cent strength, and then passed to a purification column from the head of which diethyl ether is removed, the base product being further concentrated in another still to give the constant boiling mixture of ethanol and water (95.6 per cent of ethanol by volume).

Uses of ethanol
Production of acetaldehyde (p 72)

Other compounds. These include ethyl esters, ether, chloroform and diethylaniline

Solvent. Ethanol is widely used as a solvent for a variety of organic substances, *e.g.* in varnishes, polishes, printing inks, toilet and perfumery compositions

Fig. 11. (Facing page) The ICI low pressure methanol synthesis. (By courtesy of Humphreys and Glasgow Ltd.)

Miscellaneous. Ethanol is used as a component of fuel for internal-combustion engines

C_3, C_4 and higher alcohols

Isopropanol is produced in the UK by the direct hydration (ICI) and indirect hydration of propylene. The former process is carried out at elevated temperature and pressure over a catalyst (tungsten oxide has been mentioned) but details are not available. In the indirect process propylene, under sufficient pressure to maintain it in the liquid phase, is brought into contact with 70–93 per cent sulphuric acid, the temperature being kept below 40 °C.

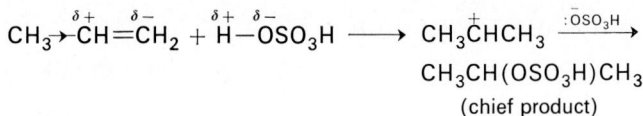

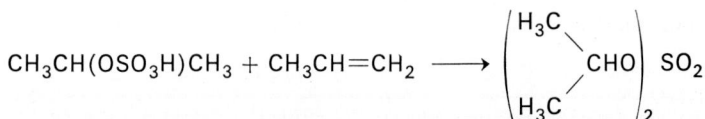

$$CH_3 \rightarrow \overset{\delta+}{CH} = \overset{\delta-}{CH_2} + \overset{\delta+}{H} - \overset{\delta-}{OSO_3H} \longrightarrow CH_3\overset{+}{C}HCH_3 \xrightarrow{:\overline{O}SO_3H}$$

$$CH_3CH(OSO_3H)CH_3$$
(chief product)

$$CH_3CH(OSO_3H)CH_3 + CH_3CH=CH_2 \longrightarrow \left(\begin{matrix} H_3C \\ \\ H_3C \end{matrix} \right\rangle CHO \right)_2 SO_2$$

The reaction mixture is then diluted with sufficient water to give a concentration of 35–50 per cent by weight of sulphuric acid, and the diluted solution is heated to bring about hydrolysis.

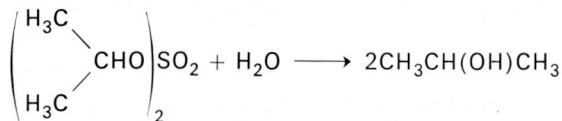

$$CH_3CH(OSO_3H)CH_3 + H_2O \longrightarrow CH_3CH(OH)CH_3 + H_2SO_4$$

$$\left(\begin{matrix} H_3C \\ \\ H_3C \end{matrix} \right\rangle CHO \right) SO_2 + H_2O \longrightarrow 2CH_3CH(OH)CH_3$$

A small amount of di-isopropyl ether is also formed by the following reaction:

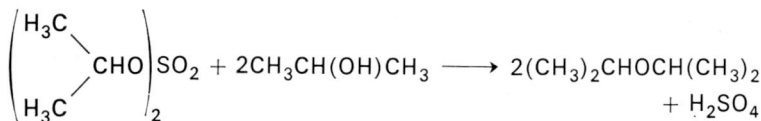

$$\left(\begin{matrix} H_3C \\ \\ H_3C \end{matrix} \right\rangle CHO \right)_2 SO_2 + 2CH_3CH(OH)CH_3 \longrightarrow 2(CH_3)_2CHOCH(CH_3)_2 + H_2SO_4$$

The above process is more widely used than the direct process in spite of disadvantages, such as large heat requirements and corrosion problems.

FIG. 12. (Facing page) Isopropanol (BP Chemicals International Ltd)—Stone and Webster Engineering Corp. (Reproduced from the November 1969 issue of *Hydrocarbon Processing*, p 193, by permission of the Gulf Publishing Co., Houston, Texas.)

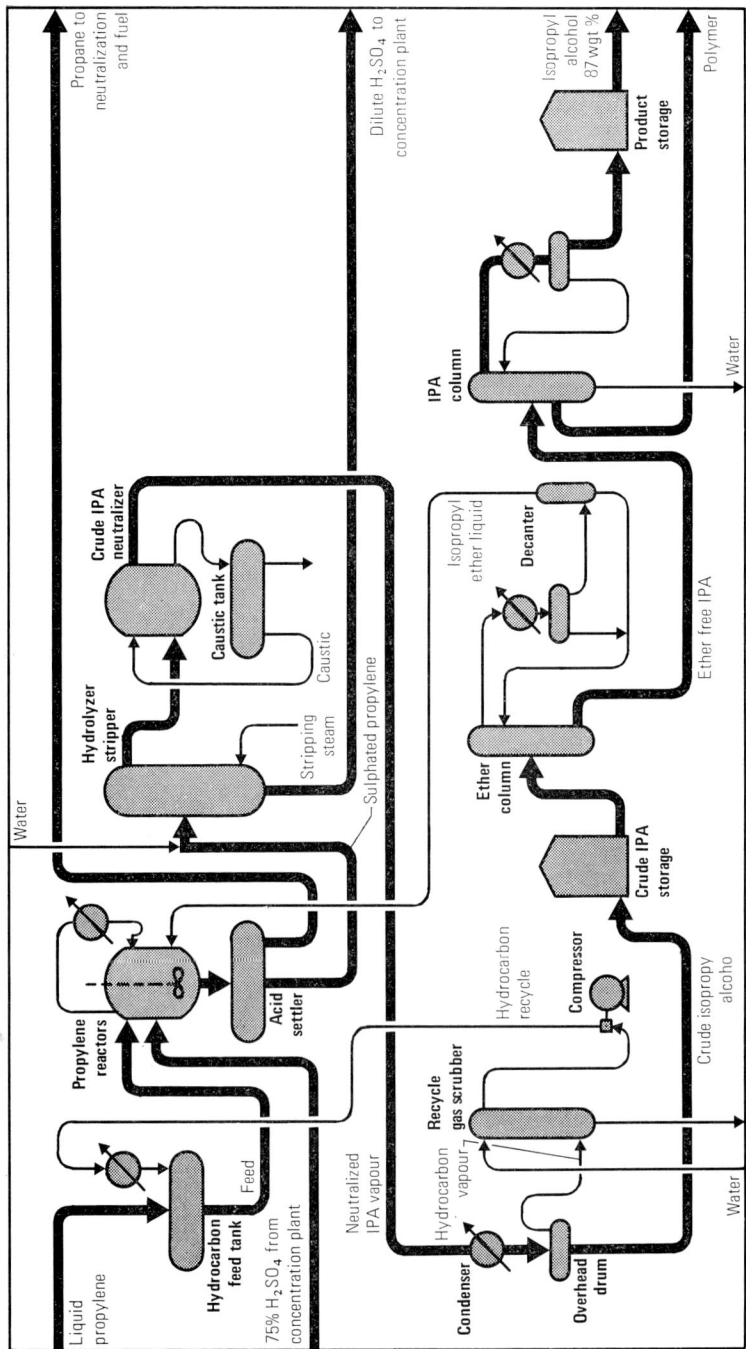

Liquid propylene

Propane to neutralization and fuel

Water

Hydrocarbon feed tank

Feed

75% H_2SO_4 from concentration plant

Neutralized IPA vapour

Water

Propylene reactors

Acid settler

Sulphated propylene

Hydrolyzer stripper

Stripping steam

Crude IPA neutralizer

Caustic tank

Caustic

Dilute H_2SO_4 to concentration plant

Hydrocarbon vapour

Condenser

Overhead drum

Recycle gas scrubber

Hydrocarbon recycle

Compressor

Water

Crude isopropyl alcoho

Crude IPA storage

Ether column

Isopropyl ether liquid

Decanter

Ether free IPA

IPA column

Water

Isopropyl alcohol 87 wgt %

Product storage

Polymer

Figure 12 is the flow diagram for the production of isopropanol by the indirect hydration of propylene (BP Chemicals International Ltd). Liquid propylene of 65 + per cent purity is employed. The absorption reaction is carried out in a series of agitated reactors at 2.026–2.375 MPa. Hydrolysis is carried out in the hydrolyser–stripper in the presence of dilution water. The reaction products are recovered by steam stripping, and the overhead vapours from the hydrolyser–stripper are neutralized with caustic soda solution and then condensed. The crude product is purified in a distillation system to give 87 per cent isopropanol. A di-isopropyl ether-rich product recovered in the distillation system is recycled. Polymer taken off from the alcohol purification column is pumped to storage. 93–95 per cent of the propylene charge is converted to isopropanol depending on the propylene content of the charge stock.

Uses of isopropanol
Production of acetone (p 75)

As a solvent for a variety of organic substances

Production of ester solvents and other compounds

The UK production of propyl alcohols in 1970 was 186 760 t.

s-Butanol is produced entirely by the indirect hydration of n-butenes, using acid of 75–85 per cent strength and keeping the temperature at 10–20 °C. It is largely used for the production of methyl ethyl ketone.

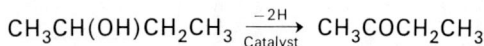

$$CH_3CH(OH)CH_2CH_3 \xrightarrow[\text{Catalyst}]{-2H} CH_3COCH_2CH_3$$

n-Butanol and isobutanol are produced largely *via* the hydroformylation of propylene.[8-10] Briefly, this is the reaction of an olefin in a solvent (if the olefin is a gas) with an equimolecular mixture of carbon monoxide and hydrogen at 90–160 °C under pressure (up to 40.52 MPa) in the presence of an oil-soluble cobalt catalyst (usually cobalt naphthenate). The actual catalyst has been shown to be cobalt carbonyl hydride $HCo(CO)_4$ $\{2Co + 8CO \rightleftharpoons [Co(CO)_4]_2 \overset{H_2}{\rightleftharpoons} 2HCo(CO)_4\}$. This has been shown to react with olefins to give aldehydes.

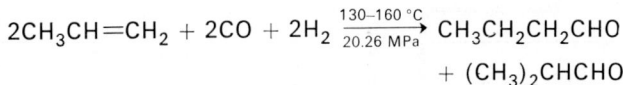

$$2CH_3CH{=}CH_2 + 2CO + 2H_2 \xrightarrow[\text{20.26 MPa}]{130{-}160\,°C} CH_3CH_2CH_2CHO + (CH_3)_2CHCHO$$

The n-isomer is produced in greater quantity than isobutyraldehyde. Hydrogenation of the aldehydes over a catalyst, *e.g.* Cu/Zn, in the vapour phase under pressure gives a mixture of n-butanol and isobutanol which are separated by fractional distillation. The process is used in the UK for the production of n-butanol and isobutanol, C_7–C_9 alcohols (from C_6–C_8 olefins) and 3,5,5-trimethylhexan-1-ol (from di-isobutenes). n-Butanol is believed still to be produced from acetaldehyde *via* aldol and crotonaldehyde.

Uses of n-butanol

Production of solvents (the most important being n-butyl acetate) and ester plasticizers (including dibutyl phthalate and butyl oleate)

As a modifying agent in the production of amino resins (from urea or melamine and formaldehyde) to produce resins used in the manufacture of stoving enamels

Uses of isobutanol

Used as a solvent in nitrocellulose lacquers and for the production of ester solvents (*e.g.* the acetate) and plasticizers

Higher alcohols

These include C_7–C_9 alcohols, 2-ethylhexanol and 3,5,5-trimethyl-hexanol, which are used largely for the production of ester plasticizers.

Polyhydric alcohols[11]

Ethylene glycol and *propylene glycol* are manufactured by heating ethylene oxide and propylene oxide (see later), respectively, with water under pressure.

Uses of ethylene glycol

As an antifreeze in car radiators

Production of Terylene. Ethylene glycol is required as a reactant with dimethyl terephthalate to produce dihydroxydiethylterephthalate, which is polymerized to Terylene (p 21)

Other compounds and miscellaneous uses. These include the dinitrate which lowers the freezing point of nitroglycerine, and glyoxal

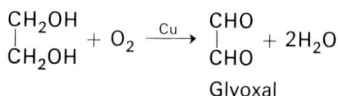

$$\begin{array}{c} CH_2OH \\ | \\ CH_2OH \end{array} + O_2 \xrightarrow{Cu} \begin{array}{c} CHO \\ | \\ CHO \end{array} + 2H_2O$$

Glyoxal

Ethylene glycol has acute and chronic toxic properties.

Uses of propylene glycol

Polyester resins. Glass-reinforced polyester resins (p 23) probably account for the greater proportion of the propylene glycol produced. It is also used as a solvent (in a pure form it is allowed in pharmaceuticals and essences on account of its non-toxicity) and in brake fluids

Glycerol[12] is obtained in the UK as a co-product in the manufacture of soaps and fatty acids from fats.[13] In the US and Holland it is also manufactured by the following process:

$$CH_3CH=CH_2 + Cl_2 \xrightarrow{450-500\ °C} ClCH_2CH=CH_2 + HCl$$

$$CH_2=CHCH_2Cl \xrightarrow[H_2O]{Cl_2} CH_2ClCH(OH)CH_2Cl \text{ (chief product)}$$
$$+ CH_2OHCHClCH_2Cl$$

$$\downarrow \begin{array}{l} Ca(OH)_2 \text{ aq} \\ -HCl \end{array}$$

$$CH_2OHCHOHCH_2OH \xleftarrow[150\ °C]{10\%\ NaOH\ aq} ClCH_2-CH-CH_2$$
$$\underset{O}{\diagdown\diagup}$$

Epichlorohydrin

In Europe synthesis only accounts for 26 per cent of the production of glycerol but in the US the figure is 46 per cent. This route is a heavy consumer of chlorine. Epichlorohydrin is much in demand as an intermediate for the production of epoxy resins, which are of value for the production of surface-coating compositions and for their electrical insulation properties. Production of epoxy resins in the UK in 1968 was 8400 t.

Glycerol is also manufactured in the US by oxidizing propylene to acrolein and reducing the latter to allyl alcohol, which is then treated with hydrogen peroxide under catalytic conditions.

$$CH_3CH=CH_2 \xrightarrow[CuO,\ 300-400\ °C]{O_2} CH_2=CHCHO$$

$$\downarrow$$

$$CH_2OHCHOHCH_2OH \xleftarrow[Catalyst]{H_2O_2} CH_2=CHCH_2OH$$

The hydrogen peroxide required for the process is made by the autoxidation of isopropanol with oxygen, with the co-production of acetone.

$$(CH_3)_2CHOH + O_2 \xrightarrow[0.3039-0.4052\ MPa]{90-140\ °C} (CH_3)_2C=O + H_2O_2$$

Uses of glycerol

Alkyd resins. These are resins used in the production of surface coatings, manufactured by the poly-condensation of phthalic anhydride with glycerol or a mixture of glycerol and pentaerythritol. UK production of alkyd resins in 1969 was 66 000 t, an increase of 2100 t on the 1968 figure

Esterification of rosin. The product, ester gum, is used as a constituent in varnishes

Nitroglycerine. This use is declining

Plasticizer for cellophane

The above constitute the major uses for glycerol. However, it is also used for many other purposes such as the production of pharmaceuticals, as a moisture conditioning agent and in foodstuffs as a preservative

References

1. E. Kilner and D. M. Samuel, *Applied organic chemistry*, p 57. London: Macdonald and Evans, 1960.
2. W. L. Faith, D. B. Keyes and R. L. Clark, *Industrial chemicals*, p 507. New York: Wiley, 1965.
3. 'Low pressure methanol synthesis', *Chem. Process.*, 1968, **4**(6), 1.
4. B. Hedley, W. Powers and R. B. Stobaugh, 'Methanol, how, where, who—future', *Hydrocarb. Process*, 1970, **49**(6), 97.
5. 'Modern wood distillation', *Chemy Ind.*, 1958, 1465.
6. T. C. Caudle and D. M. Stewart, 'Synthetic ethanol process', *Chemy Ind.*, 1962, 830.
7. Ethanol—Shell Development Co. Ltd, *Hydrocarb. Process.*, 1967, **46**(11), 168.
8. A. R. Martin, 'Hydroformylation—a new industrial route to alcohols', *Chemy Ind.*, 1956, 1400.
9. C. W. Bond, 'Synthesis of organic compounds by direct carbonylation using metal catalysts', *Chem. Rev.*, 1962, **4**(62), 283.
10. G. N. Ferguson, 'Oxo alcohols', *Chemy Ind.*, 1965, 451.
11. W. L. Faith, D. B. Keyes and R. L. Clark, *Industrial Chemicals*, p 372. New York: Wiley, 1965.
12. J. D. Thwaites, 'Glycerol—production and uses', *Chemy Ind.*, 1969, 111.
13. Ref. 1, pp 168–176.

6. Ethers and Epoxides

Diethyl ether

Diethyl ether is still produced by the continuous etherification process. In the industrial process the vapour of ethanol is passed into a heated mixture of alcohol and sulphuric acid, and acid is added from time to time to make up for losses due to side reactions.[1] Ether is also obtained as a by-product in the production of ethanol from ethylene. Ether is used as a general anaesthetic, as a solvent for the recovery of natural and synthetic products and in processes involving the use of a Grignard reagent (*e.g.* in the production of silicones).

Ethylene oxide[2,3]

Ethylene oxide was formerly produced in the UK by the dehydro-chlorination of ethylene chlorohydrin with lime. This process is believed still to be operated in the US by a firm which is a large producer of chlorine. The cost of the plant is high and the process involves the conversion of the costly chlorine into the low-value by-product calcium chloride. Ethylene oxide is usually made by the more economical direct oxidation of ethylene: using a large excess of oxygen or air at pressures up to 2.026 MPa, over a fixed-bed silver catalyst on an inert support at 200–300 °C.

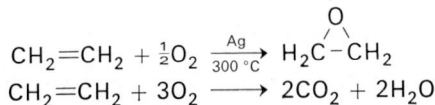

$$CH_2{=}CH_2 + \tfrac{1}{2}O_2 \xrightarrow[300\ °C]{Ag} H_2\overset{O}{\overset{\diagup\diagdown}{C}}{-}CH_2$$

$$CH_2{=}CH_2 + 3O_2 \longrightarrow 2CO_2 + 2H_2O$$

Unchanged ethylene and carbon dioxide produced in the competing reaction are recycled. Adequate removal of the heat of reaction by heat exchange between the reactor and a surrounding cooling medium is essential to reduce oxidation of ethylene and ethylene oxide to carbon dioxide and water. Production of ethylene oxide in the UK in 1970 was 174 600 t.

Figure 13 is the flow diagram for the production of ethylene oxide and ethylene glycols by the Japan Catalytic Chemical Process.

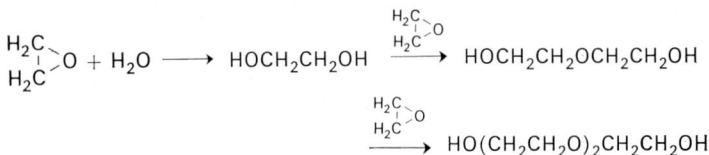

$$\underset{H_2C}{\overset{H_2C}{\diagdown}}\!\!\!\!\overset{}{\underset{}{O}} + H_2O \longrightarrow HOCH_2CH_2OH \xrightarrow{\underset{H_2C}{\overset{H_2C}{\diagdown}O}} HOCH_2CH_2OCH_2CH_2OH$$

$$\xrightarrow{\underset{H_2C}{\overset{H_2C}{\diagdown}O}} HO(CH_2CH_2O)_2CH_2CH_2OH$$

The oxidation of ethylene is carried out with air in the presence of a silver-based catalyst at about 250 °C and 2.026 MPa. High conversion efficiency is maintained by a closely controlled reactor cooling system linked with a waste heat boiler. Ethylene oxide is removed from the reactor gas by absorption in aqueous ethylene glycol. The major part of the gas stream is recycled to the main reactor while a minor proportion is diverted to a secondary reactor, absorption and energy recovery system from which inert gases are ultimately purged. Crude ethylene oxide is stripped from the glycol solution, distilled to remove water, light and heavy oils and stored under refrigeration. Some hydration of ethylene oxide occurs in the absorber solution and to prevent accumulation, part of this stream is concentrated and excess glycol removed.

Uses of ethylene oxide

Ethylene glycol (p 63). Glycol production probably accounts for more than half the ethylene oxide produced. Di- and tri-ethylene glycols are produced as by-products

Non-ionic surface-active agents. The most important non-ionic surface-active agents are those produced from ethylene oxide and long-chain alkyl phenols such as octylcresols (p 33). In 1970 European consumption of ethylene oxide in the production of non-ionic surface-active agents was 130 000 t

Other products. These include mono-, di- and tri-ethanolamines (mono-ethanolamine is an important absorbent for CO_2 and H_2S, whilst triethanol-amine is important as an emulsifying agent),

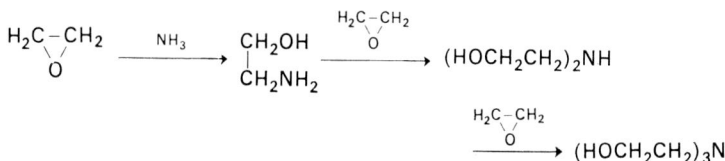

$$H_2C-CH_2 \underset{O}{\diagdown\diagup} \xrightarrow{NH_3} \begin{array}{c} CH_2OH \\ | \\ CH_2NH_2 \end{array} \xrightarrow{\underset{O}{\overset{H_2C-CH_2}{\diagdown\diagup}}} (HOCH_2CH_2)_2NH$$

$$\xrightarrow{\underset{O}{\overset{H_2C-CH_2}{\diagdown\diagup}}} (HOCH_2CH_2)_3N$$

glycol ethers (used as solvents),

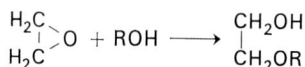

$$\begin{array}{c} H_2C \\ | \\ H_2C \end{array}\!\!>\!\!O + ROH \longrightarrow \begin{array}{c} CH_2OH \\ | \\ CH_2OR \end{array}$$

and polyethylene glycols $HOCH_2CH_2O(CH_2CH_2O-)_nCH_2CH_2OH$ produced by reaction between ethylene oxide and ethylene glycol at elevated temperature in the presence of sodium hydroxide as catalyst

Ethylene oxide is also used as a fumigant for foodstuffs and tobacco

Propylene oxide[4]

Propylene oxide is made *via* propylene chlorohydrin which is dehydrochlorinated with a slurry of lime.

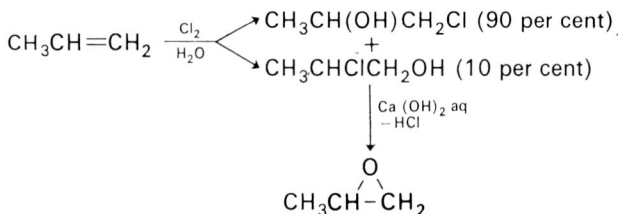

$$CH_3CH{=}CH_2 \xrightarrow[H_2O]{Cl_2} \begin{cases} CH_3CH(OH)CH_2Cl \text{ (90 per cent)} \\ + \\ CH_3CHClCH_2OH \text{ (10 per cent)} \end{cases}$$

$$\Big\downarrow \begin{array}{c} Ca\,(OH)_2 \text{ aq} \\ -HCl \end{array}$$

$$\underset{CH_3CH-CH_2}{\overset{O}{\diagup\diagdown}}$$

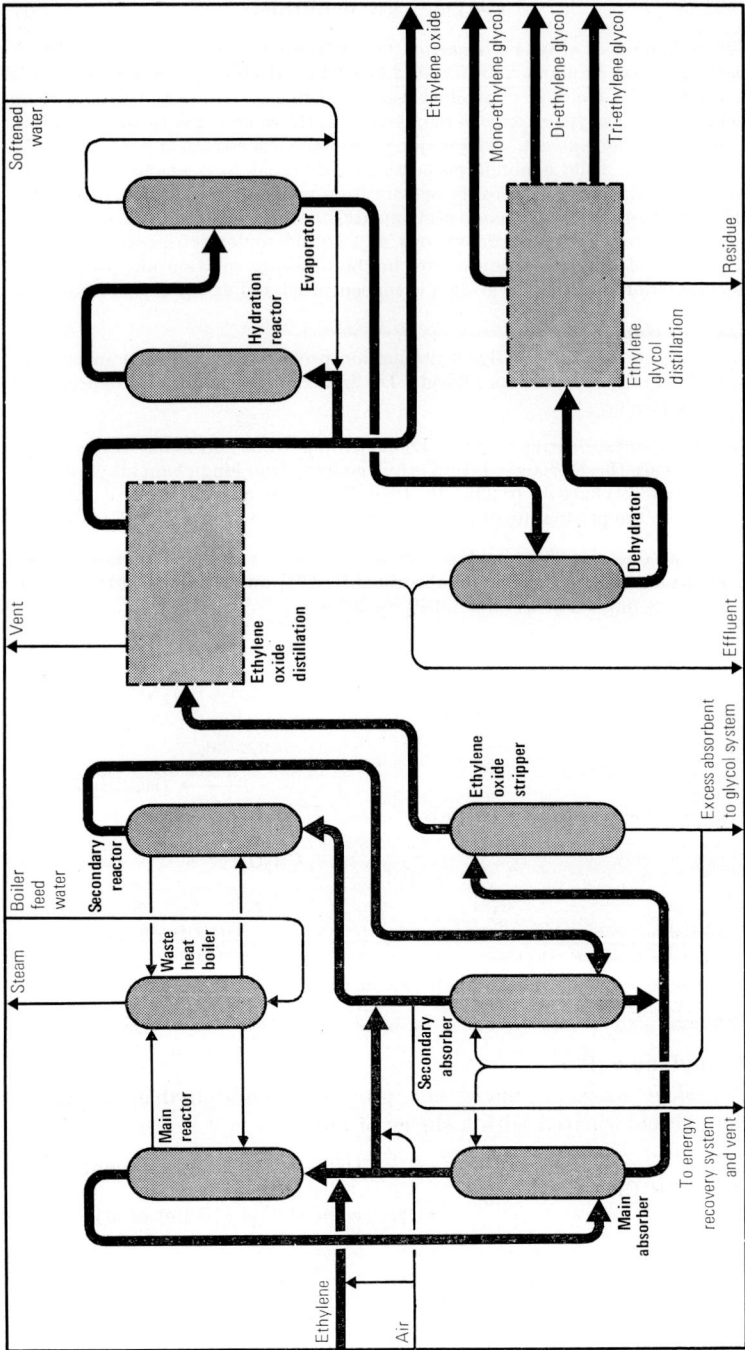

The propylene and chlorine are introduced into the base of a tower through which water is continuously circulated. The formation of propylene dichloride is minimized by employing the propylene and chlorine in a ratio of 2:1. Much research is going on to develop a process for the direct production of propylene oxide from propylene but so far only the Oxirane process has become commercial. Production of propylene oxide in the US in 1968 was 958 Mlb.

Uses of propylene oxide

Production of polyurethanes. Condensates of propylene oxide with polyhydric alcohols such as glycerol yield flexible or rigid polyurethanes on treatment with tolylene di-isocyanates. Large quantities of propylene oxide are used for this purpose. In the US this application accounts for 35 per cent of the consumption of propylene oxide

Propylene glycol (p 63):

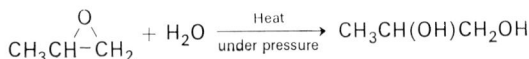

$$CH_3CH-CH_2 \overset{O}{\diagup\diagdown} + H_2O \xrightarrow[\text{under pressure}]{\text{Heat}} CH_3CH(OH)CH_2OH$$

The production of glycols accounts for about 30 per cent of the consumption of propylene oxide in the US

Other uses. Cellulose, starch and other polyhydroxy compounds on treatment with propylene oxide yield products which can be used as plastics. Co-polymers of propylene oxide and ethylene oxide are used as hydraulic fluids. The isopropanolamines, produced by reaction with aqueous ammonia, are used for some of the applications of the ethanolamines

References

1. E. Kilner and D. M. Samuel, *Applied organic chemistry*, p 74. London: Macdonald and Evans, 1960.
2. 'The manufacture of ethylene oxide and its derivatives', *Ind. Chemist*, 1963, **39**, 63.
3. Ethylene oxide, *Encyclopedia of chemical technology*, 3rd edn, (P. E. Kirk and D. F. Othmer eds), vol 5, p 304. New York: Interscience, 1964.
4. A. G. Flyvie, 'Propylene oxide and its derivatives', *Chemy Ind.*, 1964, 384.

FIG. 13. (Facing page) Ethylene oxide and ethylene glycols by the Japan Catalytic Chemicals (JCC) Process. (By courtesy of Simon–Carves Chemical Engineering Ltd.)

7. Aldehydes and Ketones

Formaldehyde[1,2]

Formaldehyde is manufactured by the air-oxidation of methanol vapour over a silver catalyst. Most processes involve dehydrogenation as well as oxidation:

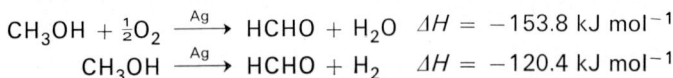

$$CH_3OH + \tfrac{1}{2}O_2 \xrightarrow{\text{Ag}} HCHO + H_2O \quad \Delta H = -153.8 \text{ kJ mol}^{-1}$$
$$CH_3OH \xrightarrow{\text{Ag}} HCHO + H_2 \quad \Delta H = -120.4 \text{ kJ mol}^{-1}$$

The catalyst is heated (electrically) to about 450 °C to start the oxidation, which is then self-supporting. The temperature in the catalyst chamber is maintained at 560–650 °C by controlling the amount of air admitted. In the US formaldehyde is also produced by the oxidation of methanol over a mixture of metallic oxides (no dehydrogenation takes place) and by the air-oxidation of propane–butane mixtures. UK production of formaldehyde in 1970 was 119 820 t (expressed as 100 per cent formaldehyde).

Figure 14 is the flow diagram for the Badische Anilin & Soda Fabrik AG formaldehyde process. Methanol and water are vaporized and mixed with compressed air. Impurities are prevented from building up in the vaporizer by a continual removal of a slip stream from the bottom of the vessel. The reaction mixture (the composition of which is above the upper ignition limit) flows through a thin layer of silver crystals where reaction takes place at slight red heat. The temperature of the catalyst is kept within ± 5 °C by automatic control of air input. The gaseous mixture leaving the reactor is cooled rapidly in a waste-heat boiler (thus preventing decomposition of formaldehyde with formation of carbon monoxide). The steam generated is sufficient to vaporize the methanol/water mixture. From the waste-heat boiler the reaction gas passes to a multi-stage absorber where it is scrubbed countercurrently with steam condensate; more than 90 per cent of the formaldehyde is absorbed in the first stage. Heat liberated during absorption and condensation is removed by heat exchangers in a product circulatory system. The concentration of product is adjusted by controlling the amount of water supplied to the last stage of the absorber. The yield is 91 per cent based on methanol converted. The catalyst (electrolytically purified silver crystals) is very selective, and one pass is sufficient to achieve almost complete conversion, making recovery of unreacted methanol by distillation unnecessary.

FIG. 14. (Facing page) Formaldehyde—Badische Anilin & Soda-Fabrik AG. (Reproduced from the November 1969 issue of *Hydrocarbon Processing*, p 187, by permission of the Gulf Publishing Co., Houston, Texas.)

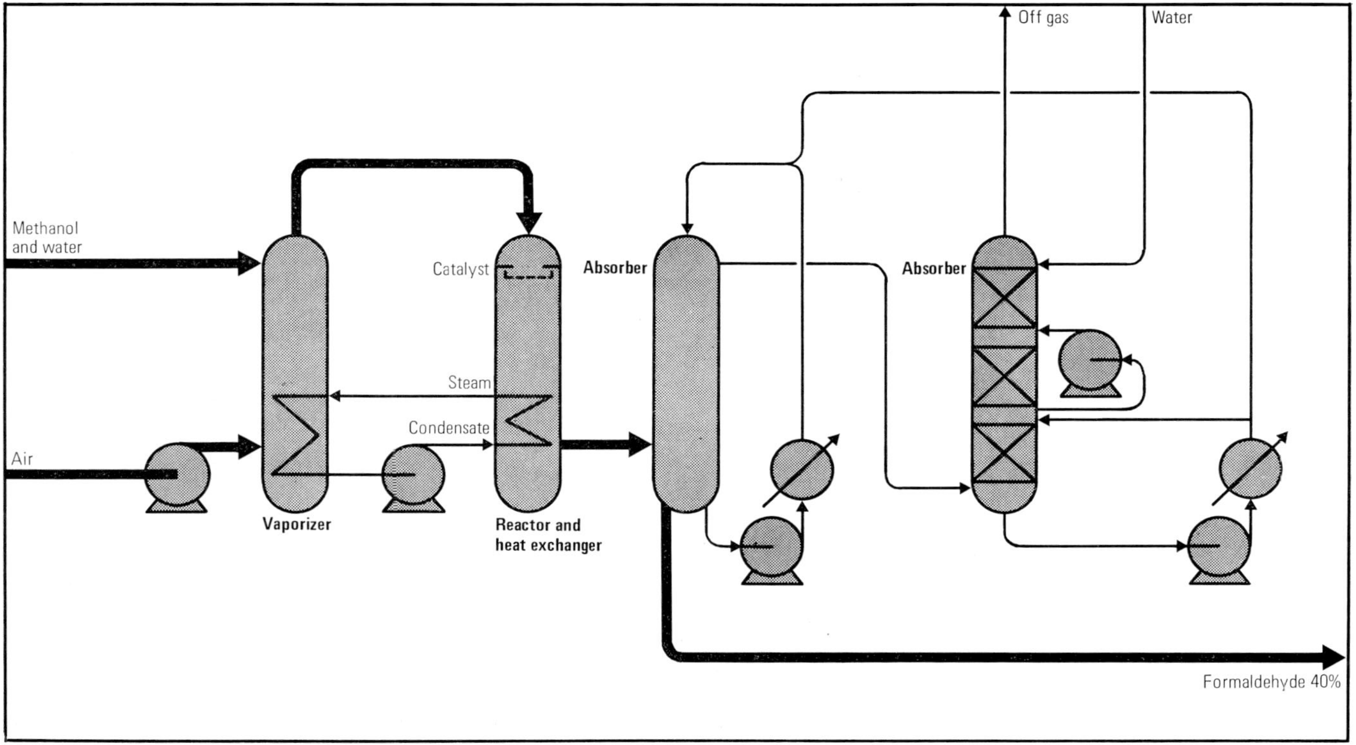

Methanol and water

Air

Vaporizer

Catalyst

Steam

Condensate

Reactor and heat exchanger

Absorber

Absorber

Off gas

Water

Formaldehyde 40%

Uses of formaldehyde

Production of phenolic resins (*i.e.* resins produced by the reaction between phenol and/or *m*- and *p*-cresols with formaldehyde) and amino-resins (from formaldehyde and urea or melamine). Phenolic and amino-resin production account for the greater part of the formaldehyde produced

Polyacetal resins.[3] In recent years polymerization of formaldehyde has been effected to give a commercially useful product. The polymerization of formaldehyde to give linear polyoxymethylenes is familiar,

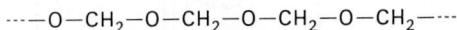

$$\cdots-O-CH_2-O-CH_2-O-CH_2-O-CH_2-\cdots$$

but such products, because of instability to heat, are of interest only as sources of formaldehyde. The problem of preparing a stable polymer was solved by Du Pont. Their product ('Delrin') is believed to be made by polymerizing highly pure anhydrous formaldehyde in a hydrocarbon solvent in the presence of an undisclosed catalyst. Depolymerization is prevented by esterifying the end hydroxyl groups and by addition of a stabilizer. Delrin is highly crystalline, softens at 175–180 °C and possesses high strength and rigidity. Its distinctive feature is its resistance to fatigue—below a certain stress it will not fail, no matter how many times stress is applied. Its combination of properties enables it to be used as a replacement for metals in light engineering applications, *e.g.* in bearings, gears and valve fittings

Other products and miscellaneous uses. These include hexamethylenetetramine (used for the production of phenolic moulding powders), trimethylolethane, $CH_3C(CH_2OH)_3$, trimethylolpropane, $CH_3CH_2C(CH_2OH)_3$, and sodium formaldehyde sulphoxylate. Formaldehyde is also used for producing washable leathers, as a hardening agent for the casein of milk to produce a plastic material used for making buttons, as a soil sterilizing agent, and for preserving zoological specimens

Acetaldehyde

Acetaldehyde is manufactured by the processes summarized in the following equations:

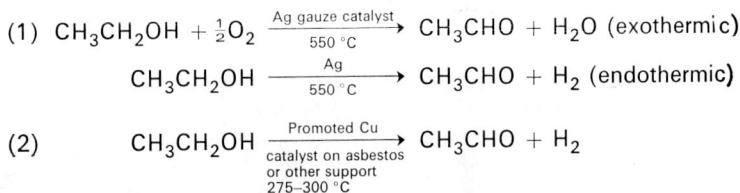

(1) $CH_3CH_2OH + \frac{1}{2}O_2 \xrightarrow[550\ °C]{\text{Ag gauze catalyst}} CH_3CHO + H_2O$ (exothermic)

$CH_3CH_2OH \xrightarrow[550\ °C]{\text{Ag}} CH_3CHO + H_2$ (endothermic)

(2) $CH_3CH_2OH \xrightarrow[\substack{\text{catalyst on asbestos} \\ \text{or other support} \\ 275–300\ °C}]{\text{Promoted Cu}} CH_3CHO + H_2$

The second process is superior to the first, being easier to control and giving a useful co-product (hydrogen).

In recent years a process has been developed in Germany for the manufacture of carbonyl compounds from olefins, in particular acetaldehyde from ethylene, acetone from propylene, and methyl ethyl ketone from 1-butene.[5] In the acetaldehyde process, ethylene and oxygen at moderate pressure are passed into a solution containing palladium chloride and copper chloride at pH 1–2:

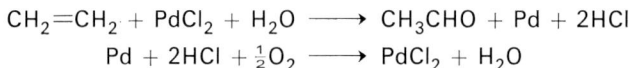

$$CH_2{=}CH_2 + PdCl_2 + H_2O \longrightarrow CH_3CHO + Pd + 2HCl$$
$$Pd + 2HCl + \tfrac{1}{2}O_2 \longrightarrow PdCl_2 + H_2O$$

The palladium chloride solution thus acts as an oxygen carrier. The purpose of the copper chloride is to promote the much slower second reaction.

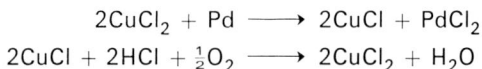

$$2CuCl_2 + Pd \longrightarrow 2CuCl + PdCl_2$$
$$2CuCl + 2HCl + \tfrac{1}{2}O_2 \longrightarrow 2CuCl_2 + H_2O$$

Thus, the overall reactions are:

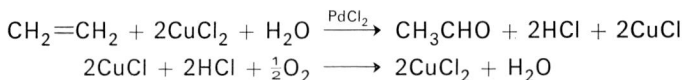

$$CH_2{=}CH_2 + 2CuCl_2 + H_2O \xrightarrow{PdCl_2} CH_3CHO + 2HCl + 2CuCl$$
$$2CuCl + 2HCl + \tfrac{1}{2}O_2 \longrightarrow 2CuCl_2 + H_2O$$

The flow diagrams for the process are shown in *Fig. 15*. Ethylene and oxygen are fed into a reactor filled with catalyst solution. The reaction takes place under slight pressure at the boiling temperature of the solution. The exothermic heat of reaction is removed by water evaporation and the concentration of the catalyst solution is kept constant by means of a corresponding water supply. The gaseous effluent leaving the reactor is scrubbed with water to recover acetaldehyde and the unreacted ethylene is recycled. Acetaldehyde is separated from by-products and water by a two-stage distillation. In the alternative two-stage process, ethylene is passed into the catalyst solution and the acetaldehyde produced is recovered by distillation. Air is then blown through the residual solution to oxidize the cuprous chloride to cupric chloride, which is then recycled.

Reference to the diminishing importance of the process for the manufacture of acetaldehyde from acetylene has been made. At present, acetaldehyde is still manufactured from alcohol in the UK. World production of acetaldehyde now exceeds 1.5 Mt/a.

Uses of acetaldehyde
Pentaerythritol

$$4HCHO + CH_3CHO + Ca(OH)_2 \longrightarrow C(CH_2OH)_4 + (HCOO)_2Ca$$

Pentaerythritol is used in the production of alkyd resins as a partial or complete replacement for glycerol

Acetic acid. The production of acetic acid by the air oxidation of acetaldehyde is still the chief outlet for acetaldehyde in many countries:

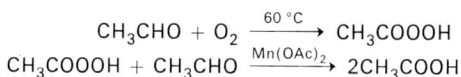

$$CH_3CHO + O_2 \xrightarrow{60\,°C} CH_3COOOH$$
$$CH_3COOOH + CH_3CHO \xrightarrow{Mn(OAc)_2} 2CH_3COOH$$

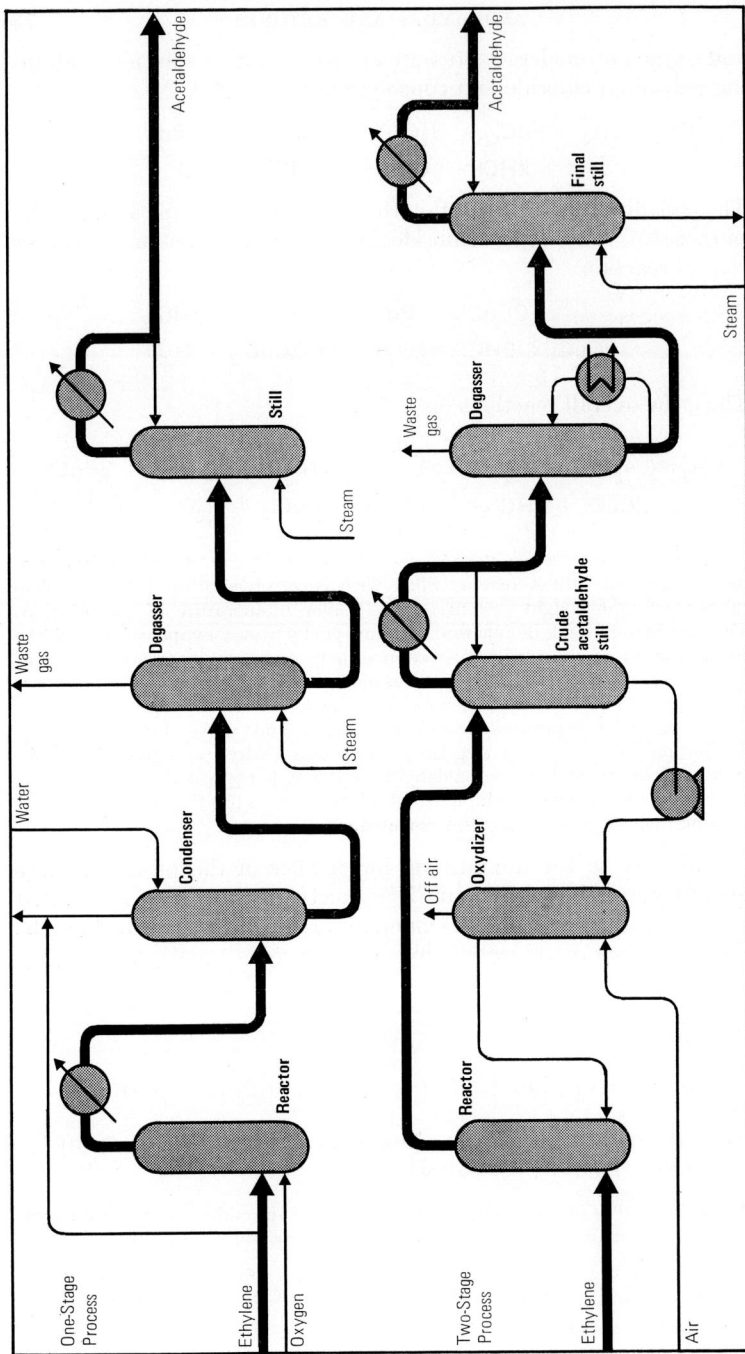

One-Stage Process

Ethylene → Oxygen → **Reactor** → **Condenser** → Water → **Degasser** → Steam → Waste gas → **Still** → Steam → Acetaldehyde

Two-Stage Process

Ethylene → Air → **Reactor** → **Oxydizer** → Off air → **Crude acetaldehyde still** → **Degasser** → Waste gas → **Final still** → Steam → Acetaldehyde

However, in the UK this process will soon be rendered obsolete since the acid is now made more economically by the oxidation of naphtha hydrocarbons

Butyraldehyde and products. Acetaldehyde was formerly important as a starting material for making n-butanol via aldol and crotonaldehyde:

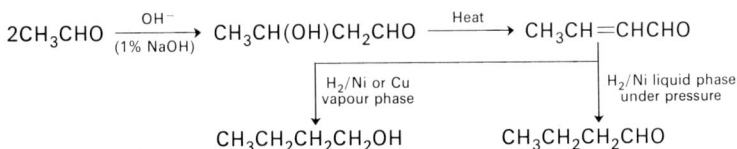

$$2CH_3CHO \xrightarrow[\text{(1% NaOH)}]{OH^-} CH_3CH(OH)CH_2CHO \xrightarrow{\text{Heat}} CH_3CH\!=\!CHCHO$$

$$\downarrow{\substack{H_2/Ni \text{ or } Cu \\ \text{vapour phase}}} \qquad\qquad \downarrow{\substack{H_2/Ni \text{ liquid phase} \\ \text{under pressure}}}$$

$$CH_3CH_2CH_2CH_2OH \qquad\qquad CH_3CH_2CH_2CHO$$

Although n-butanol is still produced by this route it is now more cheaply produced by the hydroformylation of propylene followed by hydrogenation of the n-butyraldehyde produced. Crotonaldehyde is still made by this route and n-butyraldehyde is obtained from it (by partial hydrogenation in the liquid phase, in the presence of nickel). The latter is important as the raw material for making 2-ethylhexanol

Other uses. Acetaldehyde is also used for making lactic acid (*via* acetaldehyde cyanohydrin by acid hydrolysis), trimethylolpropane, chloral (for DDT), metaldehyde (slug killer and smokeless fuel), paraldehyde (hypnotic and source of acetaldehyde, *e.g.* in the synthesis of 2-methyl-5-ethylpyridine) and vulcanization accelerators for rubber

Acetone

Acetone is manufactured by the dehydrogenation of isopropanol over copper, brass or zinc oxide at 350–380 °C:

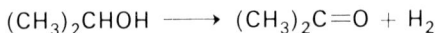

$$(CH_3)_2CHOH \longrightarrow (CH_3)_2C\!=\!O + H_2$$

and from isopropylbenzene (p 121). Acetone is also now recovered in large quantities in the UK in the course of the production of acetic acid by oxidation of naphtha hydrocarbons. It can also be obtained from propylene by oxidation in the presence of palladium chloride and copper chloride, but the process is only practical in very special circumstances. UK production of acetone in 1970 was 142 220 t.

Uses of acetone[6]
Methyl methacrylate (for Perspex)

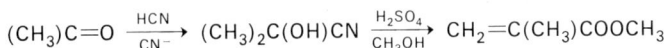

$$(CH_3)C\!=\!O \xrightarrow[\text{CN}^-]{HCN} (CH_3)_2C(OH)CN \xrightarrow[\text{CH}_3OH]{H_2SO_4} CH_2\!=\!C(CH_3)COOCH_3$$

Diacetonealcohol and derived products

FIG. 15. (Facing page) Acetaldehyde from ethylene (Aldehyd GmbH)—Hoechst-Uhde Corp. One- and two-stage processes. (Reproduced from the November 1969 issue of *Hydrocarbon Processing*, p 137, by permission of the Gulf Publishing Co., Houston, Texas.)

$$2(CH_3)_2CO \xrightarrow{\text{NaOH (s)}} (CH_3)_2C(OH)CH_2COCH_3 \xrightarrow{-H_2O}$$

$$(CH_3)_2C{=}CHCOCH_3 \xrightarrow[\text{catalyst}]{H_2} (CH_3)_2CHCH_2COCH_3 \xrightarrow[\text{catalyst}]{H_2}$$

$$(CH_3)_2CHCH_2CH(OH)CH_3$$

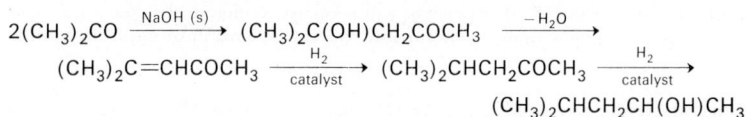

Acetic anhydride (p 79)

2,2-Bis-*p*-hydroxyphenylpropane

Solvent uses.   Acetone is widely used as a solvent, *e.g.* for cellulose acetate, in surface-coating compositions, and in paint removers

References

1. 'Formaldehyde from methanol', *Ind. Chemist*, 1960, **36**, 119.
2. W. Weimann, 'Methanol formaldehyde route boasts many benefits', *Chem. Engng*, 1970, **77**(3), 107.
3. W. H. Linton, 'Acetal polymers', *Trans. J. Plast. Inst.*, 1960, **28**(75), 131; S. J. Barker and M. B. Price, *Polyacetals*, A Plastics Institute Monograph. London: Butterworths, 1970.
4. S. A. Miller, 'Acetaldehyde manufacture—process survey', *Chem. Process. Engng*, 1968, **49**(3), 75.
5. J. Smidt, 'Oxidation of olefins with palladium chloride catalysts', *Chemy Ind.*, 1962, 54.
6. P. W. Sherwood, 'Industrial syntheses based on acetone', *Ind. Chemist*, 1956, **32**(3), 9.

8. Carboxylic Acids and Derivatives

Saturated mono-carboxylic acids

Formic acid can be manufactured by the reaction between carbon monoxide and sodium hydroxide at 200 °C under pressure. It is also obtained, (a) from its calcium salt as a co-product in the manufacture of pentaerythritol, and (b) as a by-product in the production of acetic acid (see later) by the oxidation of naphtha hydrocarbons. These latter sources probably satisfy the demand for the acid. The acid is used for the manufacture of metallic formates and esters, as a coagulant for rubber latex, as a reducing agent in the textile industry and in electroplating.

Acetic acid is by far the most important monocarboxylic acid. It was formerly manufactured largely by the oxidation of acetaldehyde in the liquid phase with air in the presence of manganese acetate as catalyst.[1] It is now manufactured in the UK by the liquid-phase air-oxidation of C_5–C_7 paraffins (from petroleum) at elevated temperature and pressure (the capacity of the plant is 90 000 t/a). The gaseous mixture leaving the reactor is cooled to condense out the products and unchanged hydrocarbons, and the condensate is separated in a distillation system into acetic acid and by-products, *viz*, formic, propionic and succinic acids, and acetone. This process will soon render the production of acetic acid from the more expensive acetaldehyde obsolete.[2]

The production of acetic acid in the US by the air-oxidation of n-butane in the liquid phase at elevated temperature and pressure in the presence of a catalyst has been described.[3] Butane is plentiful in 'wet' natural gas in the US, thus making this route economically attractive. Acetic acid is also produced (in Germany) by the non-catalytic air-oxidation of n-butane,[4] and since 1964 by the reaction between methanol and carbon monoxide in the presence of a catalyst at elevated temperature and pressure.[5]

Total UK consumption of acetic acid is 70 000–80 000 t/a.

Uses of acetic acid

Production of acetic anhydride (see later). More than half the acetic acid produced is used for making acetic anhydride

Alkyl acetates. These include acetate ester of methanol, ethanol, isopropanol, butanol and amyl alcohol. The acetates of glycerol are also important

Vinyl acetate (p 46)

Metallic acetates, including those of Al, Cu, Pb, Cr, Na, and phenylmercuric acetate

Chloroacetic acids:

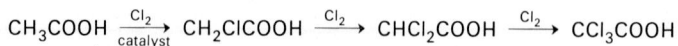

$$CH_3COOH \xrightarrow[\text{catalyst}]{Cl_2} CH_2ClCOOH \xrightarrow{Cl_2} CHCl_2COOH \xrightarrow{Cl_2} CCl_3COOH$$

FIG 16.   Production of acetic acid and by-products by the oxidation of light petroleum distillate. (By courtesy of BP Chemicals International Ltd.)

Chloroacetic acid is an important product used for the production of 2,4-dichloroacetic acid derivatives, carboxymethylcellulose, dyestuffs, and other products

Miscellaneous. Dilute solutions of acetic acid are used in the dyestuffs industry and as a coagulant for latex. Large quantities of synthetic acetic acid are used for making vinegar, for use as a preservative and flavouring agent. Acetic acid is also used as a solvent in the oxidation of p-xylene to terephthalic acid, and as a solvent and swelling agent for cellulose in the production of cellulose acetate

Acetic anhydride[6,7], $(CH_3CO)_2O$, is manufactured by several processes:

(1) From acetic acid:

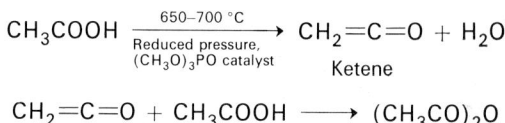

$$CH_3COOH \xrightarrow[\substack{\text{Reduced pressure,} \\ (CH_3O)_3PO \text{ catalyst}}]{650–700 \text{ °C}} CH_2{=}C{=}O + H_2O$$
$$\text{Ketene}$$

$$CH_2{=}C{=}O + CH_3COOH \longrightarrow (CH_3CO)_2O$$

The dehydration of acetic acid is carried out by passing the vapour mixed with trimethyl phosphate through heated coils. The gaseous mixture leaving the coils is cooled, the residual ketene is scrubbed with acetic acid and the mixture of anhydride and acetic acid is fractionated to separate the anhydride from the lower-boiling acetic acid.

(2) From acetone:

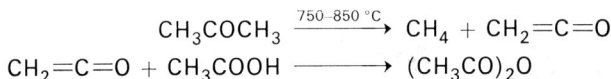

$$CH_3COCH_3 \xrightarrow{750–850 \text{ °C}} CH_4 + CH_2{=}C{=}O$$
$$CH_2{=}C{=}O + CH_3COOH \longrightarrow (CH_3CO)_2O$$

The ketene is cooled rapidly by injection of acetic acid and the mixture is then treated as described above.

(3) Other methods include one starting from acetaldehyde:

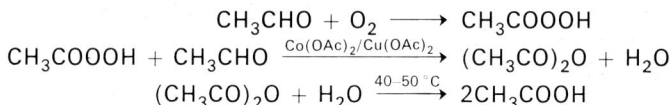

$$CH_3CHO + O_2 \longrightarrow CH_3COOOH$$
$$CH_3COOOH + CH_3CHO \xrightarrow{Co(OAc)_2/Cu(OAc)_2} (CH_3CO)_2O + H_2O$$
$$(CH_3CO)_2O + H_2O \xrightarrow{40–50 \text{ °C}} 2CH_3COOH$$

The process is carried out under conditions which minimize decomposition of the anhydride by water.

The anhydride has also been made from acetylene

Acetic anhydride is used in large quantities for the manufacture of cellulose di- and tri-acetates. The di-acetate has long been used for the production of acetate rayon textile fibre and as a plastic, whilst the tri-acetate is used for the production of a fibre, which has many of the properties of the synthetic fibres, and for making photographic film base. Cellulose acetopropionate and acetobutyrate are produced on a relatively small scale. Acetic anhydride is also used as an acetylating agent in the production of drugs such as aspirin and phenacetin.

Propionic acid is obtained in the UK as a by-product in the production of acetic acid by the oxidation of naphtha hydrocarbons. It may also be obtained by the oxidation of the corresponding aldehyde. The acid is used as a mould inhibitor in bread (largely in the form of the more convenient calcium salt) and also in the manufacture of its esters and other products.

Saturated dicarboxylic acids

Adipic acid[8,9], $HOOC(CH_2)_4COOH$, is produced in greater quantity than any other saturated dicarboxylic acid for use as an intermediate in the manufacture of nylon 66. It is manufactured by the process represented by the following equations:

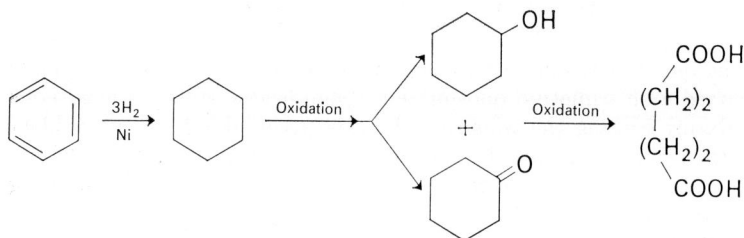

Cyclohexane can be oxidized in one stage to adipic acid, but better yields are obtained by carrying out the process in two stages. Oxidation to give a mixture of cyclohexanol and cyclohexanone is effected with air under pressure in the presence of a catalyst such as cobalt naphthenate (*see also* p 134). The resulting mixture is then fractionally distilled to recover unchanged cyclohexane, which is recycled. The second oxidation step may be effected with air or, alternatively, with 80 per cent nitric acid in the presence of a copper/vanadium catalyst at about 80 °C. In the UK adipic acid is produced from benzene but it is also made (in the US) from butadiene.

The most important use for adipic acid is as an intermediate for the manufacture of nylon 66 (p 22), but several of its esters are also important as plasticizers, particularly for pvc, and it also has other applications in the plastics industry.

Oxalic acid, $(COOH)_2 \cdot 2H_2O$, is manufactured from sodium formate but its UK production ceased in April 1966; it is now imported from Europe. The UK demand for the acid only amounts to about 600 t/a. Nevertheless, it is important as a bleaching agent and in the form of its derivatives, the diethyl ester being used as an intermediate in the manufacture of phenobarbitone and as a solvent in lacquers and printing inks. Oxalic acid has acute toxic properties.

Succinic acid, $HOOC(CH_2)_2COOH$, has been commercially available for a long time in the UK and is manufactured by the hydration of succinic anhydride, which is in turn obtained by the catalytic hydrogenation of maleic anhydride:

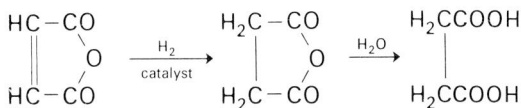

$$\underset{\substack{|| \\ HC-CO}}{HC-CO}\diagdown O \xrightarrow[\text{catalyst}]{H_2} \underset{\substack{| \\ H_2C-CO}}{H_2C-CO}\diagdown O \xrightarrow{H_2O} \underset{\substack{| \\ H_2CCOOH}}{H_2CCOOH}$$

The acid is now also obtained as a by-product in the manufacture of acetic acid by the oxidation of petroleum naphtha hydrocarbons. It is chiefly used for making pharmaceuticals, dyestuffs and resins.

Sebacic acid[11], $HOOC(CH_2)_2COOH$, is produced in fairly substantial quantities by oxidizing castor oil with concentrated caustic soda solution; capryl alcohol, $CH_3(CH_2)_5CH(OH)CH_3$, is produced as a co-product. It is important in the form of its esters, *e.g.* dioctyl sebacate, which are used as plasticizers, particularly for pvc, and as a reactant in the production of alkyd resins.

Hydroxy acids

Lactic acid, $CH_3CH(OH)COOH$, was until comparatively recently manufactured exclusively by fermentation of sugars with species of *Lactobacillus*:

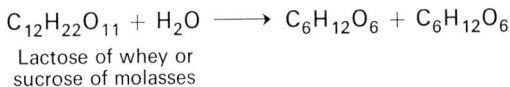

$$C_{12}H_{22}O_{11} + H_2O \longrightarrow C_6H_{12}O_6 + C_6H_{12}O_6$$
Lactose of whey or
sucrose of molasses

$$C_6H_{12}O_6 \longrightarrow 2CH_3CH(OH)COOH$$

It was produced synthetically for the first time in the US in 1963 by acid hydrolysis of acetaldehyde cyanohydrin (lactonitrile):[11]

$$CH_3CH(OH)CN + 2H_2O + HCl \longrightarrow CH_3CH(OH)COOH + NH_4Cl$$

This is a more economical method than fermentation. Hydrogen cyanide for the production of lactonitrile is now available as a by-product from the production of acrylonitrile by ammo-oxidation of propylene. About 70 per cent of the production of lactic acid is used in the food and beverage industry (in fruit sauces, pickles, carbonated beverages, dairy products and for making the dough conditioner calcium stearyl 2-lactylate). Technical lactic acid is used in the textile industry as a reducing agent, for de-liming hides, and in the form of its esters and metallic salts, the former being used as solvents and plasticizers, whilst sodium and calcium lactates have several applications in the food industry.

Citric acid[12], $HOOCCH_2C(OH)(COOH)CH_2COOH$, is largely obtained by fermenting aqueous solutions of beet molasses with the mould *Aspergillus niger*:

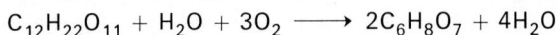

$$C_{12}H_{22}O_{11} + H_2O + 3O_2 \longrightarrow 2C_6H_8O_7 + 4H_2O$$

It is the most important acid used in the production of soft drinks, also finding application in other food products and for making metallic citrates and esters.

Tartaric acid, $HOOCCH(OH)CH(OH)COOH$, is not produced in the UK; it is imported from Europe where it is a by-product of wine production. It is not much used in the food industry, citric acid being available more cheaply. The acid is, however, used for making metallic tartrates and esters.

Unsaturated acids

Acrylic acid[13], $CH_2{=}CHCOOH$, is manufactured by several processes, including the hydrolysis of ethylene cyanohydrin (from ethylene oxide) followed by dehydration of the product, the alcoholysis of acrylonitrile, from acetylene, and by the catalytic air oxidation of propylene. The largest quantity of acrylic acid is produced from acetylene:

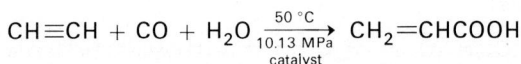

$$CH{\equiv}CH + CO + H_2O \xrightarrow[\substack{10.13 \text{ MPa} \\ \text{catalyst}}]{50\ °C} CH_2{=}CHCOOH$$

The reaction is carried out in tetrahydrofuran solution in the presence of a nickel compound as catalyst. Substitution of methanol instead of water gives the methyl ester. The ethyl ester is made by esterifying the acid with an excess of ethanol, and the butyl ester by direct esterification of the acid or from methyl acrylate by transesterification. World production of acrylic acid and acrylates in 1968 was 400 000 t. Acrylates are valuable for the readiness with which they can be co-polymerized with other monomers. The co-polymers are important in surface-coating compositions. The methyl ester is also used as a co-monomer in the production of acrylic fibres in order to increase the ability of the fibre to take up dyestuffs.

Maleic anhydride[14] is manufactured by the catalytic air oxidation of benzene:

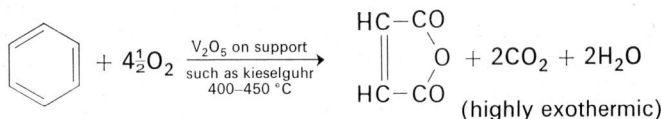

The contact time over the catalyst is 0.1 s. The process is fairly similar to the catalytic air oxidation of naphthalene and o-xylene to give phthalic anhydride (p 130). Maleic anhydride is also produced by the catalytic air oxidation of n-butenes;

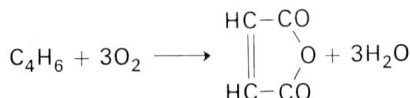

$$C_4H_6 + 3O_2 \longrightarrow \begin{array}{c} HC-CO \\ \| \quad\ \ \diagdown \\ \quad\quad\quad O \\ \| \quad\ \ \diagup \\ HC-CO \end{array} + 3H_2O$$

This process has not been used in the UK.

Uses of maleic anhydride

Glass-reinforced polyester resins. Maleic anhydride is a key component in the production of these resins which are made by preparing an unsaturated polyester from the anhydride and propylene glycol and cross-linking the product with styrene (p 25). UK production of these resins in 1969 amounted to 37 900 t. They probably take up 40–50 per cent of the maleic anhydride production

Other uses. These include the production of surface-coating resins, fumaric acid (by isomerization in aqueous solution in the presence of a catalyst), maleic hydrazide (plant growth inhibitor) and the co-polymer with methyl vinyl ether, which is used as a thickening agent. It is believed that tetrahydrofuran and y-butyrolacetone are now manufactured from maleic anhydride in the UK

Esters of carboxylic acids

The most widely used method for the production of esters is esterification of an acid by an alcohol in the presence of an acid catalyst, generally concentrated sulphuric acid (one to three per cent based on the weight of the alcohol):

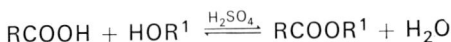

$$RCOOH + HOR^1 \xrightleftharpoons{H_2SO_4} RCOOR^1 + H_2O$$

The yield of ester is increased by displacing the equilibrium towards the right. This may be achieved by using an excess of the acid or alcohol, whichever is the more volatile (the excess reactant has to be recovered by distillation). The equilibrium is also displaced in favour of the ester by distilling off the water produced during the esterification. In esterifications involving the reaction between a lower fatty acid and ethanol or a higher alcohol, the water is removed in the form of a ternary azeotrope with the ester and excess alcohol (the ternary azeotrope boils at a lower temperature than the boiling points of the alcohol and acid). In the polycondensation of propylene glycol with a mixture of maleic anhydride and phthalic anhydride the reactants are heated under reflux in the presence of xylene, the water produced being distilled off as the binary azeotrope, xylene–water. This procedure enables the polyesterification to be carried out at a much lower temperature than in the absence of xylene. The unsaturated polyester is used for the production of glass-reinforced polyester resins. Descriptions of the manufacture of esters by the direct method have been given.[15,16]

Whilst most esters are prepared by the direct method, some are made by other methods. An important process is *transesterification* or *alcoholysis*, which is carried out in the presence of an alkaline catalyst:

$$RCOOR^1 + R^2OH \rightleftharpoons RCOOR^2 + R^1OH$$

A number of esters are manufactured by transesterification including butyl acrylate (from methyl acylate and n-butanol), dihydroxy-diethylterephthalate which on heating yields Terylene (p 21), alkyd resins, and glyceryl monostearate, which is important as an emulsifying agent in food products.[17] Vinyl acetate is prepared by an indirect method (p 46). Methyl methacrylate, the monomer used for the production of polymethyl methacrylate (Perspex), is manufactured by a process involving hydrolysis and esterification:

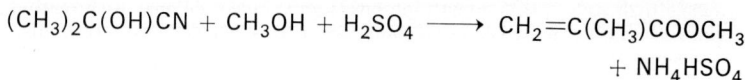

$$(CH_3)_2C(OH)CN + CH_3OH + H_2SO_4 \longrightarrow CH_2{=}C(CH_3)COOCH_3$$
$$+ NH_4HSO_4$$

Acetoacetic ester, formerly made from ethyl acetate by the Claisen condensation, is now made by treating diketene with ethanol:

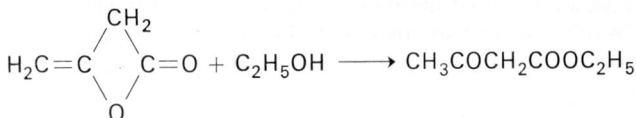

$$H_2C{=}C\underset{\diagdown O\diagup}{\overset{\diagup CH_2 \diagdown}{}}C{=}O + C_2H_5OH \longrightarrow CH_3COCH_2COOC_2H_5$$

Other esters of acetoacetic acid are made by treating diketene with the appropriate alcohol.

Uses of esters

As plasticizers. This application accounts for the largest quantity of esters. Their most important applications are for plasticizing pvc and cellulose esters. Important examples of ester plasticizers are dioctyl adipate and sebacate (for pvc) and dibutyl phthalate (for cellulose diacetate plastics)

As solvents. Esters are used as solvents in a variety of applications, *e.g.* in nitrocellulose lacquers (n-butyl acetate is particularly important for this purpose) and as extractants, *e.g.* in the manufacture of antibiotics and for the extraction of phenols from ammoniacal liquor

As flavours and in perfumery. Numerous esters are used for these purposes. Flavouring essences are used in sweets, jellies, soft drinks, *etc.*

Miscellaneous. These include the use of esters as intermediates, *e.g.* diethyl oxalate is used for making barbiturate drugs and methyl acetoacetate for making Hansa Yellow pigments. Esters containing a carbon–carbon double bond are important as monomers and co-monomers, *e.g.* methyl methacrylate and alkyl acrylates

References

1. F. J. Weymouth and A. F. Millidge, 'The manufacture and uses of acetic acid', *Chemy Ind.*, 1966, 887.
2. 'Giant acetic acid plant on stream', *Chem. Process.*, 1967, **13**(8), 4.
3. 'Acetic acid—process costs', *Chem. Process. Engng*, 1966, **47**(10), 51.
4. 'Acetic acid production from butane in W. Germany', *Chem. Trade J.*, 1964, **154**, 845.
5. 'Manufacture of acetic acid from methanol and carbon monoxide', *Chem. Trade J.*, 1965, 566.
6. E. Kilner and D. M. Samuel, *Applied organic chemistry*, p 105. London: Macdonald and Evans, 1960.
7. 'Acetic anhydride', *Encyclopedia of chemical technology*, 3rd edn, (P. E. Kirk and D. F. Othmer eds), vol 8, p 405. New York: Interscience, 1965.
8. W. L. Faith, D. B. Keyes and R. L. Clark, *Industrial chemicals*, 3rd edn, p 44. New York: Wiley, 1965.
9. Ref. 7, vol 1 (1963), p 405.
10. 'Sebacic acid', *Chem. Engng*, 1952, **59**(5), 250.
11. *Ind. Engng. Chem. ind. (int.) edn*, 1964, **56**(2), 55.
12. 'Citric acid', *Chem. Engng*, 1961, **68**(4), 22.
13. 'Acrylic acid', Ref. 7, vol 1 (1963), p 265.
14. 'Maleic anhydride should show solid growth', *Chem. Engng News*, 1968, **46**, 15.
15. E. Chadwick, 'Developments in the manufacture of organic esters', *Ind. Chemist*, 1963, **39**, 345.
16. 'Esterification', *Unit processes in organic synthesis*, 5th edn. New York: McGraw–Hill, 1958.
17. Ref. 6, 'Glyceryl monostearate', p 188.

9. Some Nitrogen Compounds

Aliphatic amines

Methylamines are manufactured by the reaction between methanol and ammonia (400 °C, 6.078 MPa) over an aluminium oxide catalyst on an inert support:

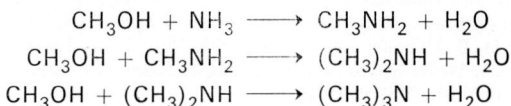

$$CH_3OH + NH_3 \longrightarrow CH_3NH_2 + H_2O$$
$$CH_3OH + CH_3NH_2 \longrightarrow (CH_3)_2NH + H_2O$$
$$CH_3OH + (CH_3)_2NH \longrightarrow (CH_3)_3N + H_2O$$

This process has been described.[1] Production of methylamines in the US in 1963 was 91 Mlb. At the moment dimethylamine is in greatest demand since it is used for the manufacture of the synthetic fibre solvents dimethylformamide and dimethylacetamide (both are

$$HCOOCH_3 + (CH_3)_2NH \longrightarrow HCON(CH_3)_2 + CH_3OH$$

also used as solvents for the recovery of butadiene by extractive distillation), unsymmetrical dimethylhydrazines (for jet and rocket fuels), agricultural and rubber chemicals, and synthetic detergents (such as lauryl dimethylamine oxide). The most important use for trimethylamine is the manufacture of choline salts (used as additives in poultry and animal feeds to stimulate growth); other uses include the manufacture of cationic surface-active agents, betaine and certain drugs. Methylamine, the least important amine, is used for the manufacture of insecticides, monomethylhydrazine, anionic surface-active agents, dyes, sarcosine and pharmaceuticals.

Ethylamines are manufactured by passing ethanol (1–1.5 parts), ammonia (6 parts) and hydrogen under pressure over a metallic hydrogenation catalyst at 150–240 °C. They have also been manufactured by the reductive amination of acetaldehyde over a nickel–chromium catalyst at elevated temperature. The latter process and the uses of the ethylamines have been described.[2] The ethylamines and the methylamines act as asphyxiant poisons, like ammonia, giving rise to inflammation of the mucous membrane linings and eyes.

Production of *ethylene diamine*[3] was started by two firms in the UK a few years ago. It is produced by heating ethylene dichloride with an excess of ammonia (concentrated aqueous or anhydrous) at elevated temperature under pressure:

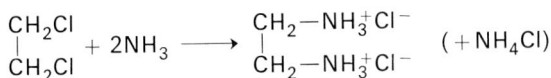

$$\begin{array}{c} CH_2Cl \\ | \\ CH_2Cl \end{array} + 2NH_3 \longrightarrow \begin{array}{c} CH_2-NH_3^+Cl^- \\ | \\ CH_2-NH_3^+Cl^- \end{array} \quad (+NH_4Cl)$$

The free base is released by addition of alkali solution and is recovered by distillation. The greater the excess of ammonia the greater the yield of ethylene diamine, but ethylene triamine, $H_2N(CH_2)_2NH(CH_2)_2NH_2$, and triethylene tetramine, $H_2N(CH_2)_2 \cdot NH(CH_2)_2NH(CH_2)_2NH_2$, are formed as by-products in appreciable quantities due to the faster rate of reaction of ethylene dichloride with ethylene diamine than with ammonia. The most important use of ethylene diamine is for the production of the well-known sequestering agent, ethylene diamine tetra-acetic acid (EDTA):

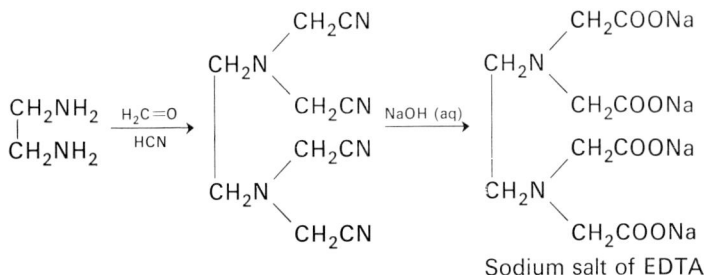

Sodium salt of EDTA

However, it is also used for making a variety of other products including basic ion-exchange resins, emulsifying agents and drugs. Ethylene diamine is an irritant substance and can give rise to dermatitis and asthma. The by-products of the above process are also in demand.

Hexamethylene diamine has already been referred to (p 22).

Acrylonitrile

Acrylonitrile was formerly made largely from acetylene.

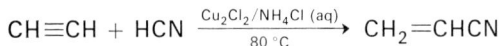

$$CH\equiv CH + HCN \xrightarrow[80\ ^\circ C]{Cu_2Cl_2/NH_4Cl\ (aq)} CH_2=CHCN$$

The acetylene and hydrogen cyanide are employed in a molecular ratio of 6–10 : 1. The acetylene route is unattractive because of the high cost of the raw materials, the stringent safety precautions necessary and operating difficulties. Most of the acrylonitrile produced is now made by the ammo-oxidation of propylene, the raw materials being relatively cheap compared with those used in the acetylene process.[4,5] The process consists of passing a mixture of propylene, ammonia, air and steam over a catalyst (*e.g.*, bismuth molybdate or phosphomolybdate on silica) at about 450 °C (contact time over the catalyst is about 2 s):

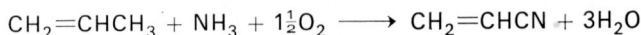

$$CH_2{=}CHCH_3 + NH_3 + 1\tfrac{1}{2}O_2 \longrightarrow CH_2{=}CHCN + 3H_2O$$

The process undoubtedly involves the intermediate formation of acrolein,

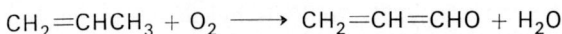

$$CH_2{=}CHCH_3 + O_2 \longrightarrow CH_2{=}CH{=}CHO + H_2O$$

which then undergoes ammo-oxidation to give acrylonitrile.

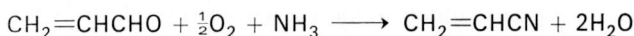

$$CH_2{=}CHCHO + \tfrac{1}{2}O_2 + NH_3 \longrightarrow CH_2{=}CHCN + 2H_2O$$

Two by-products are formed: hydrogen cyanide (up to 0.1 t per ton of acrylonitrile) and a small quantity of acetonitrile. Both are readily disposed of.

Figure 17 is the flow diagram for the Sohio process. Approximately stoichiometric quantities of propylene (85 + per cent), ammonia (fertilizer grade) and air are introduced into a fluid-bed catalytic reactor operating at 33–202.6 kPa and 399–510 °C. The reactor effluent is scrubbed in countercurrent with water and the organic materials are recovered from the absorber water by distillation. The crude product is fractionated to remove hydrogen cyanide, water, light ends and high-boiling impurities, producing fibre-grade acrylonitrile. The original catalyst used in the process is based principally on molybdenum and bismuth. A new catalyst produces more acrylonitrile and less by-product acetonitrile. A high conversion is obtained on a once-through basis in the fluid bed reactor. Troublesome separation and recycling of unchanged raw materials is not necessary. Yields in excess of 0.80 lb of acrylonitrile per pound of propylene are achieved. About 0.15–0.20 lb of by-product hydrogen cyanide can be recovered per pound of acrylonitrile.

Uses of acrylonitrile

Acrylic fibres. These are co-polymers produced from acrylonitrile and a basic monomer (2-vinylpyridine is often mentioned), the latter serving to give the co-polymer sufficient basic character to make it receptive to acid dyestuffs. Acrylic fibres are spun from solution in a solvent such as dimethylformamide. They account for 80–90 per cent of the UK production of acrylonitrile. World acrylic fibre production now exceeds 600 000 t/a

Butadiene–acrylonitrile co-polymer rubbers (p 30)

FIG. 17. (Facing page) Acrylonitrile (Sohio Process).—The Badger Co. Inc. (Reproduced from the November 1969 issue of *Hydrocarbon Processing*, p 146, by permission of the Gulf Publishing Co., Houston, Texas.)

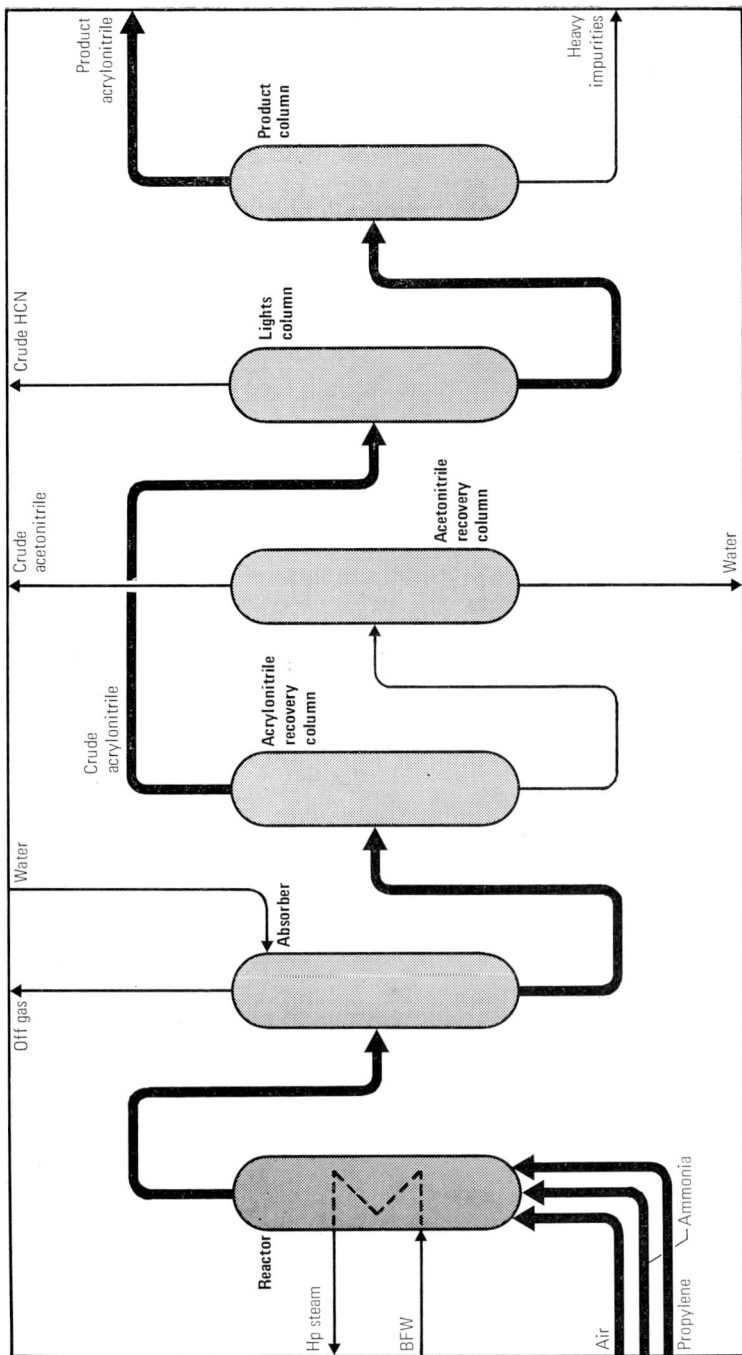

Reactor

Hp steam

BFW

Air

Propylene

Ammonia

Off gas

Water

Absorber

Crude acrylonitrile

Acrylonitrile recovery column

Crude acetonitrile

Acetonitrile recovery column

Crude HCN

Water

Lights column

Heavy impurities

Product column

Product acrylonitrile

Plastics. ABS polymers have grown rapidly in importance during the last five years or so. They consist of dispersions of polybutadiene rubber, grafted with styrene and acrylonitrile in a matrix of styrene–acrylonitrile co-polymer. They are outstanding for their resistance to fracture by impact over a wide range of temperatures and also possess a good strength-to-weight ratio. They are invaluable for applications where components are subjected to much handling and rough treatment. In the US in 1969, 90 Mlb of ABS polymers were used by the motor car industry. UK production of ABS polymers in 1968 amounted to 5000 t

Other uses. Acrylonitrile is also used in the US for the production of adiponitrile (an intermediate in the manufacture of nylon 66) by an electrolytic process:

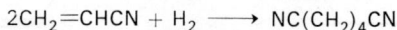

$$2CH_2{=}CHCN + H_2 \longrightarrow NC(CH_2)_4CN$$

Acrylonitrile is a very toxic substance and even in low concentrations can give rise to toxic symptoms including flushing, nausea and shallow breathing (it prevents oxygen being absorbed by the body cells).

Urea

Urea is manufactured in very large quantities in the UK (figures not available). US production in 1968 amounted to 4856 Mlb. Even in 1966 world production amounted to about 10 Mt/a. Urea is manufactured by the reaction between ammonia and carbon dioxide at elevated temperature and pressure.[6,7]

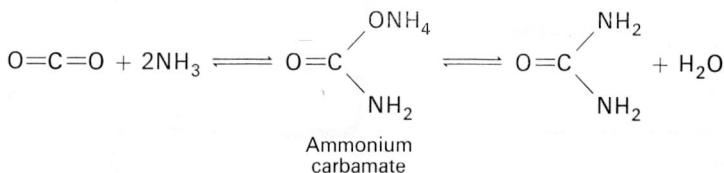

Ammonium
carbamate

The percentage conversion of carbamate to urea (a slow process) depends on the amount of excess ammonia employed. The various urea processes differ largely in the ways in which unchanged carbon dioxide and ammonia are dealt with. Urea solution can be produced by a *once-through process,* a *partial recycle process,* or a *complete recycle process.*[8] In the once-through process (*Fig. 18*) gaseous carbon dioxide and liquid ammonia are fed to the reactor at a mole ratio of about 2.5 : 1. Under the operating conditions (21.58 MPa and 185 °C) about 60 per cent of the carbon dioxide is converted to urea. The heat of reaction is used to produce steam in the reactor coils. The reactor effluent consisting of urea, unchanged carbamate, water and excess ammonia, at 0.1013 MPa is passed through a steam-heated decomposer where the unconverted carbamate is completely decomposed to carbon dioxide and ammonia. These gases are

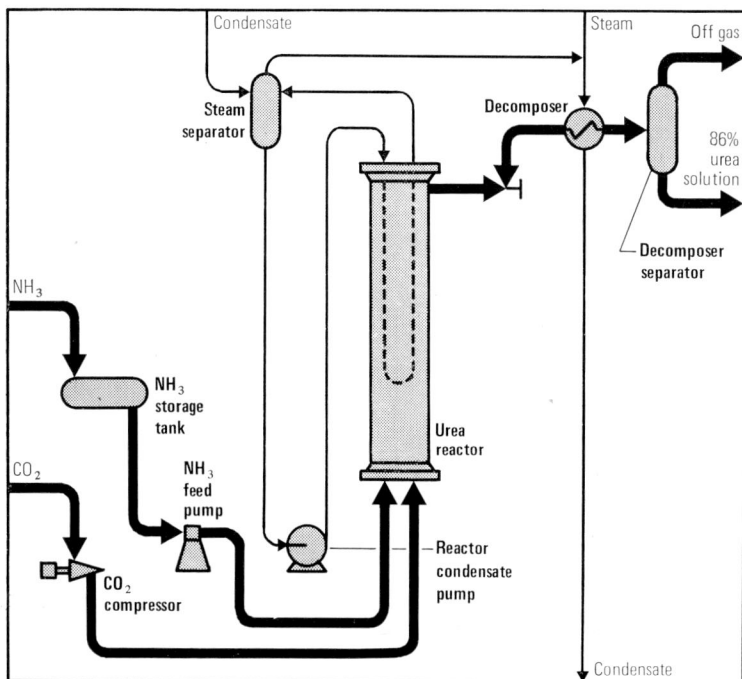

FIG. 18. Production of urea by the once-through process. (By courtesy of the Chemical Construction Corp.)

separated from the urea solution in the decomposer–separator and can be used in plants for the production of ammonium nitrate, ammonium phosphate or ammonium sulphate.

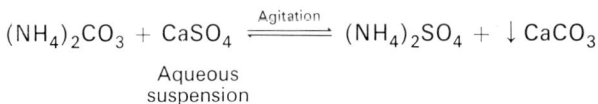

$$(NH_4)_2CO_3 + CaSO_4 \xrightleftharpoons{\text{Agitation}} (NH_4)_2SO_4 + \downarrow CaCO_3$$

Aqueous
suspension

The 86 per cent urea solution is processed to yield prilled urea by evaporation prilling (*Fig. 19*), atmospheric crystallization and prilling or vacuum crystallization and prilling. Crystals can be obtained from the last two processes. In the partial recycle and complete recycle processes, carbon dioxide and ammonia are fed to the reactor at a mole ratio of four parts of ammonia to one part of carbon dioxide and about 70 per cent of the carbon dioxide is converted to urea.

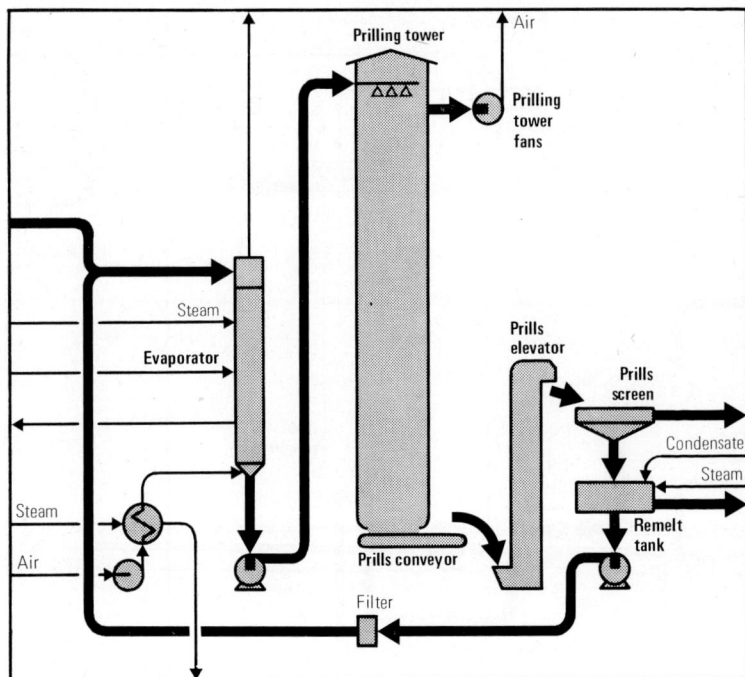

Fig. 19. Urea-finishing process—evaporation prilling. (By courtesy of the Chemical Construction Corp.)

Uses of urea

Urea-formaldehyde resins. In the UK production of these resins uses up more urea than is consumed as fertilizer. Production of amino resins (*i.e.* urea–formaldehyde and melamine–formaldehyde resins) in the UK in 1969 was 134 100 t, an increase of about 7000 t on the 1968 figure

Fertilizer. Urea is the most concentrated of the straight fertilizers, containing 46 per cent of nitrogen. It is exported in large quantities from the UK for use in tropical climates, where it is greatly in demand. It is being increasingly used in granular compound fertilizers because of its very high nitrogen content

Miscellaneous. These include the isolation of n-paraffins (p 37) and the production of barbiturate drugs and chloroisocyanuric acid

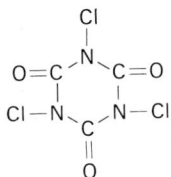

Chloroisocyanuric acid

The latter is used in applications requiring active chlorine for bleaching, *e.g.* scouring powders

An important new use for urea is for the manufacture of *melamine* (for melamine–formaldehyde resins). In the UK melamine is made from calcium cyanamide by a well-established process. In the urea process the latter is heated strongly to give a mixture of cyanic acid and ammonia which is then passed over a catalyst at elevated temperature:[9]

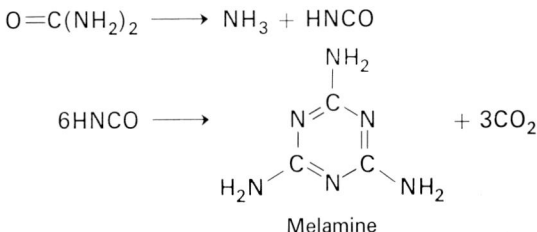

$$O=C(NH_2)_2 \longrightarrow NH_3 + HNCO$$

$$6HNCO \longrightarrow \text{Melamine} + 3CO_2$$

Melamine

The production of calcium cyanamide from calcium carbide for melamine production is declining; the urea process is more economical since the latter is available at low cost.

References

1. R. Williams, J. R. Willmer and J. Schaefer, *Chem. Engng News*, 1955, **33**, 3982.
2. 'Ethylamines', *Encyclopedia of chemical technology*, 3rd edn, (P. E. Kirk and D. F. Othmer eds), vol 1, p 122. New York: Interscience, 1963.
3. 'Ethylene diamine', Ref. 2, vol 7 (1965), p 349.
4. 'Acrylonitrile—latest synthesis for propylene', *Chem. Engng*, 1965, **72**(3), 150.
5. 'Acrylonitrile—process costs', *Chem. Process. Engng*, 1967, **48**(6), 855.
6. W. L. Faith, D. B. Keyes and R. L. Clark, *Industrial chemicals*, 3rd edn, p 790. New York: Wiley, 1965.
7. A. J. Payne, 'Urea—process survey', *Chem. Process. Engng*, 1969, **49**(5), 81
8. Urea—Bulletin 102. 'Chemico plants, processes and experience'. Chemical Construction Corporation, 320 Park Avenue, New York, N.Y. 10022.
9. 'A new route to melamine from urea', *Chem. Engng*, 1965, **72**(10), 180.

10. Hydrocarbons

Production of aromatic compounds from coal[1-3]

Carbonization of coking coal is at present carried out in the UK for the following purposes: the production of metallurgical coke, in coke ovens at about 1300 °C, and organic by-products consisting of gas, tar and crude benzole (*i.e.* the product recovered by scrubbing the gas with gas oil after removal of tar and ammonia); the production of gas in continuous vertical retorts at 900–1000 °C, the by-products consisting of coke, tar and crude benzole; and the production of the smokeless semi-cokes Coalite and Rexco in intermittent vertical retorts at 550–600 °C, the by-products consisting of gas, tar and crude gas spirit. In addition, the NCB produces two smokeless fuels by fluid-bed carbonization of bituminous coal.

The tar and crude benzole obtained by coke oven carbonization contain a higher percentage of aromatic hydrocarbons than the corresponding products obtained by carbonization in continuous vertical retorts, but tar acids (phenols) are present in greater quantity in gas works tar. These differences in composition arise since coke ovens are operated at higher temperatures than continuous vertical retorts. Crude benzole derived from coke ovens has a much higher aromatic hydrocarbon content than that derived from gas works (which may contain up to 25 per cent of non-aromatic hydrocarbons). For this reason the gas works product has been used mainly for the production of motor benzole and solvents, whereas crude benzole from coke ovens is an important source of pure aromatic hydrocarbons.

Carbonization of coal in gas works is declining and is expected to cease in a few years time as natural gas from the North Sea becomes more available. The amount of tar and crude benzole available from coke ovens is governed by the demand for coke for smelting iron ore.[4] Although steel production is likely to increase over the next five years or so, coke production is likely to remain static since the amount of coke required for smelting purposes is being progressively reduced as a result of improvements in the smelting process (in 1950 in the UK about 20 cwt of coke were required for the production of one ton of pig iron, whereas in 1964 the figure was only 13.9 cwt). Production of crude benzole in the UK in 1969 from gas works and coke oven works amounted to 7.7 and 79.2 Mgall, respectively.

(The estimated 1972 production figures for crude benzole are 1.2 and 74 Mgall, respectively.) The quantity of crude benzole available from coke ovens is likely to account for only about 20 per cent of the needs of the chemical industry for benzole products by the late 1970's.[5]

Carbonization of one ton of coal in a coke oven yields only about 0.01 t of crude benzole. The crude benzole may contain more than 70 per cent of benzene, up to 15–16 per cent of toluene and 3–5 per cent of xylenes, together with other aromatic compounds. Light oil obtained by fractional distillation of coal-tar also contains benzene as the main component and is combined with crude benzole for refining. It is usual first to wash with 10 per cent aqueous sodium hydroxide solution to remove phenols and then with 25 per cent sulphuric acid to remove tar bases (pyridine, *etc.*). Carbon disulphide and other low-boiling compounds are then removed as a forerunnings fraction by steam distillation. The residue from the steam distillation is fractionally distilled to give a mixture of benzene, toluene and xylenes, and a higher-boiling fraction which is used mainly as a solvent (heavy solvent naphtha). The mixture of benzene, toluene and xylenes is then agitated with a small amount of concentrated sulphuric acid, which dissolves sulphur compounds and unsaturated hydrocarbons capable of polymerization, washed first with water and then with sodium hydroxide solution, and finally fractionally distilled to give benzene, toluene and mixed xylenes, the latter being sold without separation for use as a solvent. An alternative process to acid refining is *hydrorefining*; the vapour of the crude benzole (which is first subjected to treatment to remove tar acids and bases) is mixed with hydrogen (present in coke oven gas) and the mixture passed under pressure over a catalyst (*e.g.* cobalt molybdate) at about 350 °C, under which conditions aromatic hydrocarbons are not hydrogenated but sulphur-containing compounds react, *e.g.*

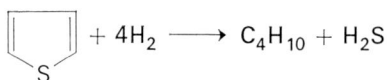

Thiophen

$$CS_2 + 4H_2 \longrightarrow CH_4 + 2H_2S$$

The plant required is much more expensive than that required for acid refining, but the product, as would be expected, has a much lower sulphur content.

The UK production of crude tar (from gas works, coke oven works and low-temperature carbonization processes) in 1969 was 1.67 Mt, but by 1972 production will probably have fallen to 1.3–

1.4 Mt.   The production of gas works tar, which in 1966 amounted to 1.2 Mt, will probably only amount to about 70 000 t in 1972 and by 1975 production is expected to cease.   The major uses of coal-tar are for road-surfacing and in the form of creosote–pitch fuel which is widely used in steel works.   However, it is also important as a source of certain organic compounds.   The tar is separated from the bulk of the water present by decanting or by centrifuging, residual water being removed by distillation.   It is then separated by fractional distillation in continuous pipe stills into five fractions and a pitch residue.   The first fraction (final bp about 170 °C), *light oil*, is refined with crude benzole.   The second fraction, *carbolic oil*, has a boiling range of 170–200 °C and is rich in tar acids.   The third fraction boils over the range 200–230 °C, is the source of naphthalene, but also contains tar acids and bases.   The fourth fraction, *creosote oil*, is not further processed.   The fifth fraction, boiling between 270 °C and 360 °C, is the only commercial source of anthracene, but it also contains carbazole (which is recovered), phenanthrene and other polynuclear compounds.

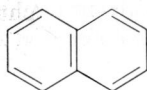   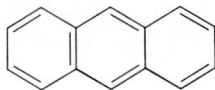   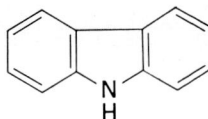

   Naphthalene   Anthracene   Carbazole

   Phenanthrene

Tar acids are recovered by washing the second and third fractions from coal-tar with 10 per cent alkali solution (the extracts are combined with a similar extract from the washing of crude benzole and light oil),

$$ArOH + NaOH \rightleftharpoons ArONa + H_2O$$

and decomposing the extract by contacting it with carbon dioxide in a packed tower:

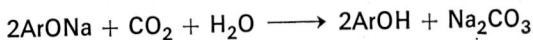

$$2ArONa + CO_2 + H_2O \longrightarrow 2ArOH + Na_2CO_3$$

The crude tar acid mixture is then dehydrated by distillation and separated by fractional distillation to give phenol, *o*-cresol, *m*- and *p*-cresol mixture, xylenols and higher-boiling tar acids.   A good deal of the mixture of tar acids is sold under the name Cresylic Acid without further separation. Production of refined Cresylic Acid in the UK in 1967 was 13.5 Mgall. The most important source of tar acids has been gas works tar which in a few years time will no longer be available. The loss of cresols and xylenols from this source can be made good by increasing production from low-temperature tar (which is very rich in tar-acids) and from coke oven tar, and also by new synthetic processes.

The following tar bases are recovered from the acid extracts already referred to:

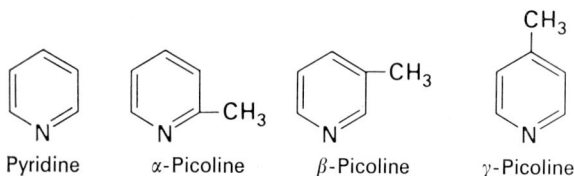

Pyridine   α-Picoline   β-Picoline   γ-Picoline

Pyridine bases are also synthesized,[6] including α-picoline and 2-methyl-5-ethylpyridine.

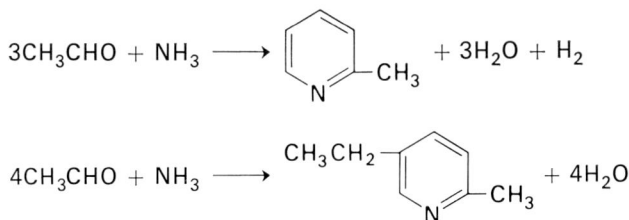

$$3CH_3CHO + NH_3 \longrightarrow \text{[picoline]} + 3H_2O + H_2$$

$$4CH_3CHO + NH_3 \longrightarrow \text{[ethylmethylpyridine]} + 4H_2O$$

These two compounds are used for producing 2-vinylpyridine and 2-methyl-5-vinylpyridine, respectively. Production of pyridine bases by synthesis is now greater than the quantity recovered from the products of coal carbonization (969 000 gall in 1967). Pyridine itself is in considerable demand for the production of the bipyridylium weed killers, Diquat and Paraquat (ICI Ltd), sales of which are believed to exceed £8 m/a, but it has not yet been produced

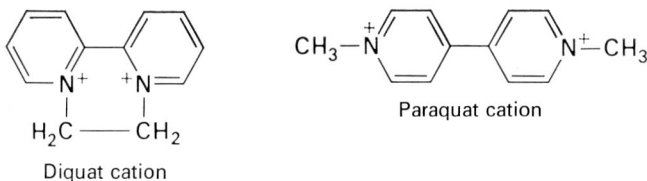

Paraquat cation

Diquat cation

synthetically by an economical method. However, the increased demand can be met by subjecting alkylpyridines to hydrodealkylation (*see* p 102).

Naphthalene is recovered from the neutral oil which results after the naphthalene oil fraction has been washed with alkali and acid solutions. The neutral oil is cooled and the resulting cake of crystals is purified by crystallization or by centrifuging, the former process giving a purer product. Production of naphthalene (crude and refined) in the UK in 1967 was 84 600 t. Anthracene is recovered from the fifth fraction from coal-tar by allowing the latter to cool and recovering the crystals formed by centrifuging or by vacuum filtration. The product so obtained contains about 20 per cent of anthracene and is upgraded to 40–50 per cent anthracene by treatment with solvent naphtha (which dissolves out other compounds) followed by filtration. On average only 3500–4000 t/a of anthracene are recovered, there being no demand for greater production, although much more is available in the tar.

Low-temperature carbonization, established in the UK in 1927, is increasing in importance because of the increasing demand for smokeless semi-cokes in the domestic market, whilst the by-products are sufficiently important to be regarded as co-products. It now accounts for 5–8 per cent of the coal carbonized in the UK. The amount of coal carbonized by the Coalite process[7,8] is approaching 2 Mt/a and the semi-coke produced accounts for about 10 per cent of the UK smokeless fuel market. Carbonization is carried out at 550–600 °C in intermittent vertical retorts. About 16.5 gall of tar are produced for every ton of coal carbonized by the Coalite process. The tar is much richer in tar acids than high-temperature tar and is becoming increasingly important as a source of tar acids. The product corresponding to crude benzole is called crude gas spirit and differs from the former in that it contains only about 25 per cent of aromatic hydrocarbons; it is refined for use as motor spirit. About 40 gall of aqueous liquor are produced for every ton of coal carbonized, from which monohydric and dihydric phenols are recovered by extraction with isobutyl acetate.

Production of aromatic hydrocarbons from petroleum[9]

During the last few years there has been a very great increase in the quantity of aromatic hydrocarbons produced from petroleum in the UK. Production capacity by four oil companies and by ICI, is now probably approaching two Mt/a. It has been estimated that the production of benzene from petroleum will be five times greater that that produced from coal by 1975. It would be impracticable to recover the small amounts of aromatic hydrocarbons present in crude oils, since their boiling points are near to those of the accompanying paraffins and cycloparaffins. The chief source of aromatic hydrocarbons is now petroleum naphtha, isolated by fractional distillation of crude oil and having a maximum bp of about 170 °C. It consists of paraffins, naphthenes and aromatic hydrocarbons. In naphtha derived from Middle East crude oil paraffins predominate, whereas in Nigerian crude oils paraffins and naphthenes are present in about equal proportions, aromatic hydrocarbons being present in only small amounts (7–12 per cent). Aromatic hydrocarbons are produced from naphtha by catalytic reforming, the most widely used process being *platforming* (Universal Oil Products)[10] which takes place over a catalyst consisting of 0.3 to 0.7 per cent platinum on silica–alumina. The naphtha fraction is first re-fractionated to remove hydrocarbons lower than C_6 and higher than C_9 or C_{10}. It is then de-sulphurized if its sulphur content is too high, by subjecting it to a mild hydrogenation process (*hydrorefining*) over a suitable catalyst, such as activated cobalt and molybdenum oxides on alumina, at elevated temperature, when organic sulphur is removed as hydrogen sulphide. The vaporized naphtha is then mixed with recycle hydrogen (which suppresses deposition of coke on the catalyst, thus ensuring its continued activity) and the mixture is passed under pressure (up to 1.317 MPa) through a reactor containing the catalyst at 450–530 °C. The effluent from the reactor is cooled and the gaseous material (mostly hydrogen) is separated from the liquid products and partly recycled (after compression). In practice, hydrogen is produced in excess of the amount needed in the process and can be used for catalytic hydrogenation processes. The liquid product may contain from 40–60 per cent of aromatic hydrocarbons. It is stabilized (*i.e.* fractionally distilled at about 1.317 MPa) to remove hydrogen and low-boiling hydrocarbons (largely propane and butanes). The stabilized platformate can either be used as high-octane blending stock for motor spirit or as feedstock for the production of pure aromatic hydrocarbons.

Figure 20 illustrates the Esso powerforming plant. In practice, the naphtha vapour is passed through four furnaces and reactors in series. The catalyst gradually becomes deactivated and must, therefore, be regenerated. The first powerformer at the Fawley Refinery of the Esso Petroleum Co. Ltd operates on a cyclic principle—a 5th reactor, known as the 'swing' reactor, enables each of the four operating reactors to be taken out of service in succession for catalyst regeneration. In the case of the new Fawley powerformer which came into operation in 1968 there is no swing reactor; the whole unit is shut down periodically and the reactors are regenerated simultaneously.

Aromatic hydrocarbons are produced in the course of the reforming process by the following reactions:

(1) Dehydrogenation of naphthenes, *e.g.*

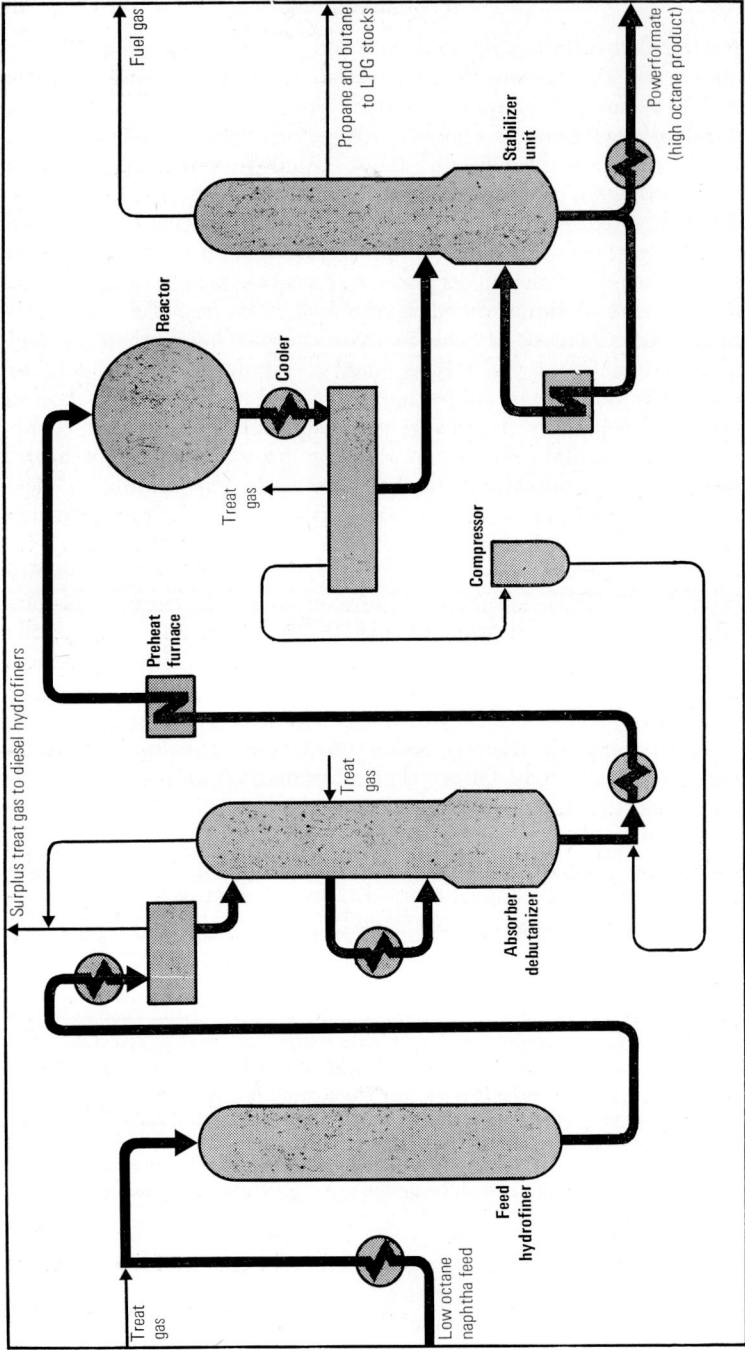

Fuel gas

Propane and butane
to LPG stocks

Powerformate
(high octane product)

Stabilizer
unit

Reactor

Cooler

Treat gas

Compressor

Preheat
furnace

Surplus treat gas to diesel hydrofiners

Treat gas

Absorber
debutanizer

Feed
hydrofiner

Treat gas

Low octane naphtha feed

$$+ 3H_2$$

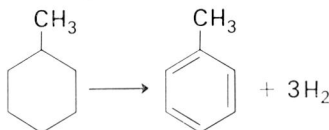

(2) Cyclization of paraffins followed by dehydrogenation:

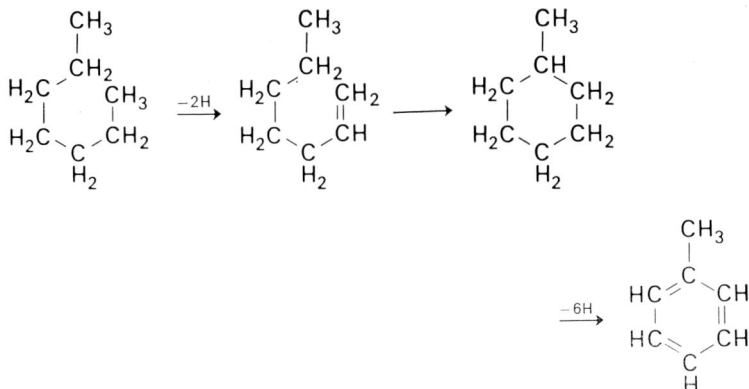

Amongst other reactions which occur in the course of reforming that may be mentioned are isomerization of paraffins and naphthenes and hydrocracking of paraffins, *e.g.*

$$(CH_3)_2CH(CH_2)_3CH_3 + H_2 \longrightarrow (CH_3)_3CH + CH_3CH_2CH_3$$

The most widely used process for the separation of the aromatic hydrocarbons present in reformate from the accompanying non-aromatic hydrocarbons (largely paraffins) is counter-current extraction of the aromatics with a suitable selective polar solvent.[11] Formerly, the most widely used solvent was diethylene glycol and 8–10 per cent of its weight of water, but now more selective solvents are used, the following being particularly important:

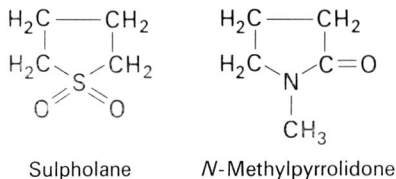

Sulpholane   *N*-Methylpyrrolidone

Sulpholane is now being used by several firms in the UK. The non-aromatic hydrocarbons can be incorporated into motor spirit or used as fuel. The extract is then fractionally distilled to separate

FIG. 20. (Facing page) Powerforming. (By courtesy of Esso Petroleum Co. Ltd.)

the aromatic hydrocarbons from the solvent, which is recycled. The aromatic hydrocarbons are separated by fractional distillation in a series of columns to give benzene, toluene and C_8 aromatics, leaving a residue which, in the US, is a source of durene (tetramethyl-benzene) and is also fractionated to provide feedstock for the production of naphthalene by hydrodealkylation (see later).

It is important to note that catalytic reformate contains benzene, toluene and xylenes, in increasing order of quantity, whereas benzene is present in much greater quantity than toluene and the xylenes in crude benzole derived from coal carbonization processes. The C_8 fraction contains o-, m- and p-xylenes and ethylbenzene,

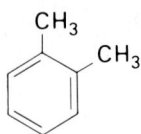

o-Xylene,
bp 144.5 °C,
fp −25.3 °C

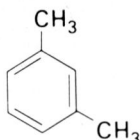

m-Xylene,
bp 139.2 °C,
fp −47.9 °C

p-Xylene,
bp 138.4 °C,
fp 13.3 °C

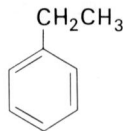

Ethylbenzene,
bp 136 °C

and is separated into its components by a combination of fractional distillation and low-temperature fractional crystallization (p-xylene has a much higher freezing point than o- and m-xylenes). The yield of p-xylene can be increased by subjecting the residue remaining after the recovery of p-xylene (the demand for m-xylene in the UK is small) to vapour phase isomerization over a catalyst such as platinum ('octafining' process) in the presence of hydrogen. Ethylbenzene (if not previously recovered) is unaffected by the isomerization. In recent years additional supplies of o- and p-xylenes have become available by the disproportionation of toluene.

Hydrodealkylation[12,13]

Toluene is available in far greater quantity than benzene from catalytic reforming processes, but the demand for the latter as a feedstock for the chemical industry is much larger than for toluene. Consequently, processes have been developed for producing benzene from toluene. Toluene vapour is mixed with compressed hydrogen (in the catalytic process steam may be introduced to minimize deposition of carbon on the catalyst), and the mixture is passed under pressure through a reactor (at about 550 °C) which may or may not contain a catalyst. Molybdenum compounds (on alumina) are said to be suitable catalysts. The chief reaction which occurs is hydro-dealkylation

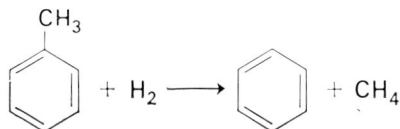

but side-reactions also lead to methane and diphenyl:

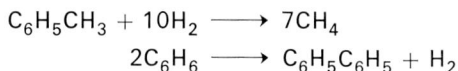

$$C_6H_5CH_3 + 10H_2 \longrightarrow 7CH_4$$
$$2C_6H_6 \longrightarrow C_6H_5C_6H_5 + H_2$$

The process is also used in the US for the production of naphthalene from methylnaphthalenes present in the residual fraction from catalytic reforming already referred to:

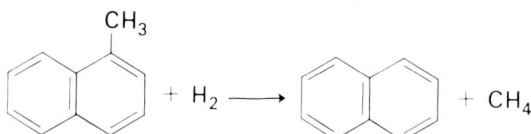

Petroleum sources now account for about 40 per cent of the naphthalene produced in the US but this figure is not likely to increase.

Figure 21 is the flow diagram for a thermal hydrodealkylation process. Hydrogen-rich gas is compressed and mixed with toluene. The mixture is preheated (heat exchange and in a fixed preheater) and charged to the reactor. The exothermic heat of reaction is removed by intermediate quenches when necessary. The mixture leaving the reactor is quenched, cooled by heat exchange with reactor feed and finally by cooling water. It then passes to a gas/liquid separator. Some of the gas may be recycled and the remainder removed for use as fuel gas. The liquid product is charged to a stabilizer column from the head of which fuel gas is removed. The stabilizer product is treated with clay and fractionated to yield pure benzene and recycle toluene (and accompanying diphenyls). The yield of benzene is greater than 99 per cent of the theoretical, whilst that of naphthalene from methylnaphthalenes is expected to be in excess of 95 per cent of the theoretical yield.

Uses of aromatic hydrocarbons
UK production of benzene in 1970 was 445 790 t

Benzene, toluene and the xylenes (in reformate and motor benzole) are important ingredients in motor spirit, owing to their high octane numbers

Uses of benzene in the chemical industry
Ethylbenzene. Probably about one-third of the benzene available is used for this purpose. Nearly all the ethylbenzene produced is used for making styrene

Cyclohexane.

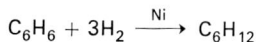

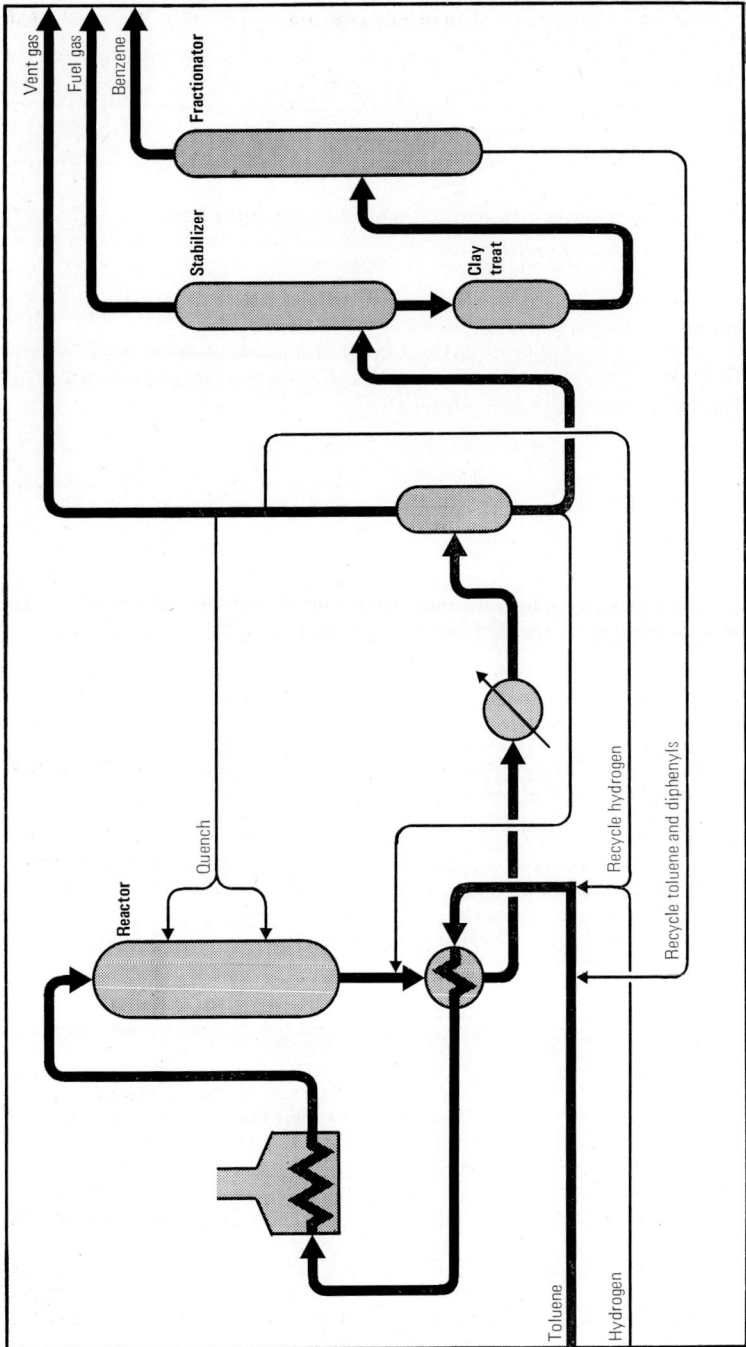

About 90 per cent of the cyclohexane produced is used for the production of nylons 6 and 66. Cyclohexane now accounts for about 64 per cent of the raw materials used in Europe for the manufacture of caprolactam (which is polymerized to nylon 6). The production of cyclohexane probably accounts for 25–30 per cent of the benzene available in the UK

Phenol (p 120)

Anionic surface-active agents (p 27)

Substances produced in lesser quantities
Maleic anhydride (p 80)

Chlorobenzene[14] (for the production of DDT and nitrochlorobenzenes):

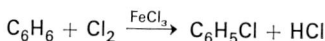

$$C_6H_6 + Cl_2 \xrightarrow{FeCl_3} C_6H_5Cl + HCl$$

Polychlorobenzenes. *o*-Dichlorobenzene is used as a dyestuff intermediate; *p*-dichlorobenzene is important as a moth repellent

Hexachlorocyclohexanes:[15]

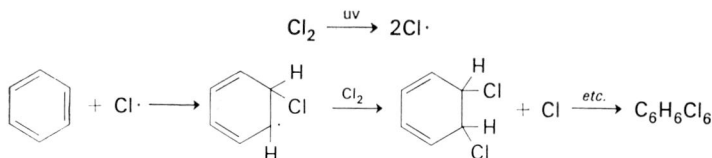

$$Cl_2 \xrightarrow{uv} 2Cl\cdot$$

$$+ Cl \xrightarrow{etc.} C_6H_6Cl_6$$

The process is carried out in a glass-lined vessel fitted with an agitator.

The γ-isomer which, unlike the other accompanying isomers, has insecticidal properties, is present to the extent of 13–14 per cent. The process is not allowed to go to completion and after removal of some of the unchanged benzene the α- and β-isomers are precipitated by crystallization

Aniline (p 115)

Biphenyl:

$$2C_6H_6 \xrightarrow[\text{temperature}]{\text{Elevated}} C_6H_5C_6H_5 + H_2$$

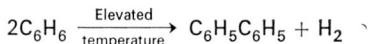

Biphenyl with diphenyl ether is used as a heat transfer agent whilst polychlorodiphenyls are used as components of industrial paints and as plasticizers

Toxicity of aromatic hydrocarbons
Benzene is toxic by absorption through the skin and by inhalation of the vapour. Chronic poisoning gives rise to anaemia and can also

FIG. 21. (Facing page) Hydrodealkylation—Atlantic Richfield Co., Hydrocarbon Research, Inc. (Reproduced from the November 1969 issue of *Hydrocarbon Processing*, p 187, by permission of the Gulf Publishing Co., Houston, Texas.)

produce a predisposition to leukemia.[16]   In the laboratory it should only be used under fume-cupboard conditions and if possible toluene or, better still, white spirit substituted. Toluene and the xylenes are considered to be less hazardous and chronic poisoning due to these substances is believed to be comparatively rare.   However, they have acute narcotic and toxic properties, the maximum allowable concentration in air being 200 ppm.

Uses of toluene in the chemical industry[17]

The UK production of toluene in 1970 was 207 830 t. No new large scale uses for toluene as a chemical feedstock are visualized in the near future. If the amount of lead tetraethyl added to motor spirit is reduced, toluene will increase in importance since it has a higher octane number than the other common aromatic hydrocarbons, and additional quantities will be required in motor spirit to offset the drop in octane number. In addition to its uses as a chemical feedstock it is used as a solvent in surface-coating compositions, particularly in nitrocellulose lacquers

In the US the largest single use for toluene is for the production of benzene by hydrodealkylation, but the most important UK uses are for the production of 2,4- and 2,6-tolylene di-isocyanates (p 113) and TNT

In the US toluene is also important as the starting material for the production of phenol by the Dow process (p 123) whilst in Italy it serves as the starting material for the production of nylon 6

Formerly toluene was used as the starting material for benzoic acid production by chlorination to benzotrichloride and hydrolysis of the latter. Benzoic acid is, however, now made by direct catalytic air oxidation of toluene (p 128). Chlorination of toluene at its bp to give benzyl chloride and benzyl-idene chloride[18] is still important in the UK, the former being used for the production of benzyl alcohol and the latter for the production of benzalde-hyde

Toluene is also used for the production of o-toluenesulphonyl chloride (for saccharin manufacture) and of dyestuff and pigment intermediates, including the nitrotoluenes, toluidines and nitrotoluidines (especially the pigment intermediate, 3-nitro-4-aminotoluene)

Uses of o-xylene[19]

Most of the o-xylene available is used for the production of phthalic anhydride (p 130). By 1980 o-xylene will probably account for about 75 per cent of the UK production of phthalic anhydride

A minor use is the production of 4-amino-o-xylene, used for making vitamin B_2

Uses of m-xylene[19]

In the US large quantities of m-xylene are used for the production of isophthalic acid by catalytic air oxidation, the latter in turn being used for making alkyd resins and plasticizer esters

Minor uses of m-xylene include the production of dyestuff and pigment intermediates and the local anaesthetic xylocaine, important in dentistry *m*-Xylene is also important as a solvent

Uses of p-xylene[19]

By far the most important use for *p*-xylene is the production of terephthalic acid which is used as such or more commonly in the form of its dimethyl ester for the production of the polyester fibre, Terylene

Minor uses include the production of dyestuff intermediates and pesticides. 1970 production of *p*-xylene in the US is expected to be about 1.8 Mlb.

Uses of naphthalene[20]

Phthalic anhydride (p 130). About 75 per cent of the naphthalene produced in the UK is oxidized to phthalic anhydride. However, competition from *o*-xylene derived from petroleum as a raw material is increasing and it is expected that an appreciable surplus of naphthalene will eventually be available for export

Dyestuff and pigment intermediates. The intermediate produced in greatest quantity is 2-naphthol but several others are important, including 2-hydroxy-3-naphthoic acid and various aminosulphonic acids

Miscellaneous uses. These include the production of insecticides, rubber antioxidants, 1-naphthylacetic acid (which prevents premature dropping of apples and pears), additives for fuels and lubricating oils, perfumery ingredients and polychloronaphthalenes

Production of ethylbenzene and styrene[21,22]

Styrene, $C_6H_5CH{=}CH_2$, is made by a two-stage process from benzene:

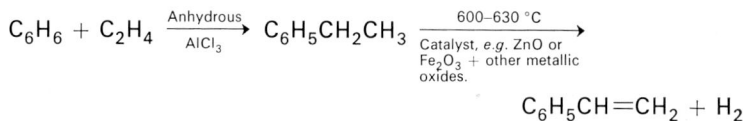

$$C_6H_6 + C_2H_4 \xrightarrow[\text{AlCl}_3]{\text{Anhydrous}} C_6H_5CH_2CH_3 \xrightarrow[\substack{\text{Catalyst, } e.g. \text{ ZnO or} \\ \text{Fe}_2\text{O}_3 + \text{other metallic} \\ \text{oxides.}}]{600\text{--}630\ °C}$$

$$C_6H_5CH{=}CH_2 + H_2$$

The alkylation process is carried out at 90–100 °C and at atmospheric pressure or slightly above. The benzene is present in the reaction mixture in excess of the theoretical quantity in order to minimize the formation of the accompanying polyethylbenzenes. A possible mechanism for the alkylation is as follows:

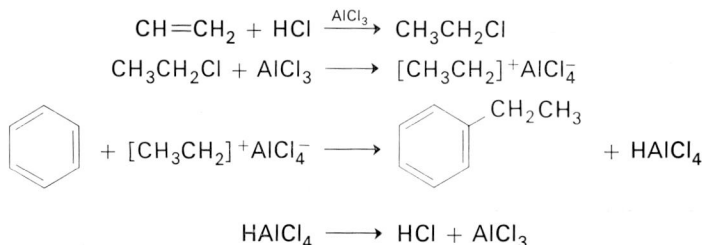

$$CH{=}CH_2 + HCl \xrightarrow{\text{AlCl}_3} CH_3CH_2Cl$$

$$CH_3CH_2Cl + AlCl_3 \longrightarrow [CH_3CH_2]^+AlCl_4^-$$

$$HAlCl_4 \longrightarrow HCl + AlCl_3$$

It is thus necessary to add a small quantity of either ethyl chloride or hydrogen chloride. The process is carried out continuously in a glass or enamel-lined reaction vessel fitted with an agitator and a reflux condenser. The contents of the vessel are withdrawn continuously to a separating vessel where the crude reaction product is separated from a lower layer of complex formed between aluminium chloride, benzene and ethylated benzenes, which is returned to the reaction vessel. The crude ethylbenzene is washed with water followed by alkali solution and fractionally distilled in a series of three columns. From the head of the first column unchanged benzene is taken off and is recycled, from the head of the second column ethylbenzene is taken off, and from the third column polyalkylbenzenes are distilled and recycled to the reaction vessel where they undergo de-ethylation in the presence of aluminium chloride:

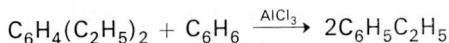

$$C_6H_4(C_2H_5)_2 + C_6H_6 \xrightarrow{AlCl_3} 2C_6H_5C_2H_5$$

Figure 22 is a photograph of a high-yield process operated by the Sinclair–Koppers Co., producing ethylbenzene by the alkylation of benzene with ethylene in the presence of aluminium chloride. The benzene-make-up and recycle-benzene stream are combined and dried by distillation prior to their introduction into the reactor. A small quantity of ethyl chloride is added to the ethylene feed to act as promoter.

FIG. 22. Production of ethylbenzene by the alkylation of benzene with ethylene. (By courtesy of the Badger Co. and the Sinclair–Koppers Co.)

The dehydrogenation of ethylbenzene is carried out in the presence

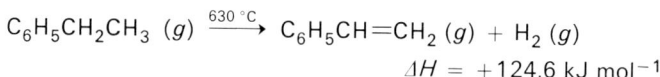

$$C_6H_5CH_2CH_3 \ (g) \xrightarrow{\text{630 °C}} C_6H_5CH{=}CH_2 \ (g) + H_2 \ (g)$$
$$\Delta H = +124.6 \text{ kJ mol}^{-1}$$

of a large excess of superheated steam which reduces the partial pressure of the ethylbenzene to a few kPa (the dehydrogenation involves an increase in volume and is therefore favoured by low pressure) and acts as a source of heat. The conversion per pass is 40–60 per cent. The styrene is recovered by fractional distillation under vacuum, the unchanged ethylbenzene being recycled. A polymerization inhibitor for styrene is added to the crude product before distillation. The inhibitor, p-t-butylcatechol, is also added to the distilled product.

Uses of ethylbenzene and styrene

Most of the ethylbenzene available is used for making styrene. A small amount is used for the production of acetophenone by catalytic air oxidation

Production of plastics and synthetic rubbers. 1968 production of styrene in the US was 3698 Mlb. 1970 consumption of styrene in Western Europe is expected to exceed 1 Mt

Polystyrene. Polystyrene production in the UK in 1969 amounted to 145 700 t, an increase of 14 000 t on the 1968 figure. It accounts for about 50 per cent of the styrene produced. The polymer is available as the unmodified material, as blown or expanded polystyrene (widely used for packaging) and as toughened polystyrene. The latter is produced by blending polystyrene with butadiene-styrene co-polymer rubber (about 5 per cent of the latter considerably increasing the impact strength and elongation)

Co-polymers. Styrene is co-polymerized with butadiene to produce the general purpose rubber SB-R, and high-styrene rubbers which are particularly important in the footwear trade

ABS polymers are outstanding for their high impact strength (p 88)

Styrene is also a key compound in the production of glass-reinforced polyester resins, being used as a cross-linking agent to give a thermosetting product

Exposure to concentrations of styrene vapour greater than 100 ppm is not advisable.

Finally, another important aromatic hydrocarbon is cumene (isopropylbenzene). It is produced by alkylating benzene with propylene in the presence of anhydrous aluminium chloride as catalyst and is used as an intermediate for the production of phenol and acetone (p 121).

References

1. D. M. Samuel, 'Coal and petroleum as raw materials for the production of aromatic compounds', *Chemy Ind.*, 1968, 567.
2. H. G. Franck, 'The challenge in coal-tar chemicals', *Ind. Engng Chem. ind. (int.) Edn*, 1963, **55**(5), 38.
3. A. Bradley, 'Chemicals from coal', *Chemy Ind.*, 1965, 2024.
4. 'Future trends in the development of the coking industry', *Chemy Ind.*, 1963, 1060.
5. M. Edwards, 'Benzole, its future in relation to the chemical industry', *Chem. Age*, 1967, 97, No. 2500, 15.
6. D. McNeil, 'Pyridine from novelty to leader', a new chemistry supplement to the October 1967 issue of *Chem. Process.*
7. J. Gubb, 'Chemicals from coal—low temperature carbonization', *Chem. Brit.*, 1968, **4**(7), 291.
8. J. G. M. Thorne, 'Coal chemicals—by-products no longer', *Chem. Process.*, 1970, **16**(3), 25.
9. See ref. 1.
10. D. A. Williams and G. Jones, *Liquid fuels, platforming*, p 14. Oxford: Pergamon, 1961.
11. See ref. 1.
12. G. F. Asselin, 'Hydrodealkylation', in *Advances in petroleum chemistry and refining*, vol 9, p 4. New York: Interscience, 1964.
13. 'Hydrodealkylation processes', *Ind. Engng Chem. ind. (int.) Edn*, 1962, **54**(2), 28.
14. W. L. Faith, D. B. Keyes and R. L. Clark, *Industrial chemicals*, 3rd edn, p 261. New York: Wiley, 1965.
15. 'Benzenehexachloride manufacture—process costs', *Ind. Engng Chem. ind. (int.) Edn*, 1956, **48**(10), 41A.
16. 'Benzene, toluene and nicotine as industrial hazards', *Chem. Age*, 1956, April 7th, 799.
17. R. B. Stobaugh, 'Toluene, how, where, who—future', *Hydrocarb. Process.*, 1966, **45**(2), 139.
18. 'Benzyl chloride, benzal chloride and benzotrichloride', *Encyclopedia of chemical technology*, 3rd edn, (P. E. Kirk and D. F. Othmer eds), vol 5, p 28. New York: Interscience, 1964.
19. R. B. Stobaugh, 'Xylenes, how, where, who—future', *Hydrocarb. Process.*, 1966, **45**(4), 149.
20. R. B. Stobaugh, 'Naphthalene, how, where, who—future', *Hydrocarb. Process.*, 1966, **45**(3), 149.
21. J. N. Hornibrook, 'Manufacture of styrene', *Chemy Ind.*, 1962, 872.
22. S. A. Miller and J. W. Donaldson, 'Styrene manufacture—process costs', *Chem. Process. Engng*, 1967, **48**(12), 37.

11. Some Aromatic Nitro- and Amino-Compounds

Nitrobenzene[1,2]

The manufacture of nitrobenzene would be better illustrated on a laboratory scale if it was carried out in a vessel fitted with an electrically driven stirrer instead of in a test-tube. On the large scale a cast-iron reaction vessel fitted with an agitator and a thermometer is used; cooling is achieved by circulating water through lead cooling coils inside the vessel (ΔH for the nitration of benzene with mixed acid is $-112.9\,\mathrm{kJ\,mol^{-1}}$). Mixed acid (approximate composition: 57 per cent H_2SO_4, 35 per cent HNO_3 and 8 per cent H_2O) is added gradually to the stirred benzene, the temperature being maintained at 50–55 °C. When the nitration is complete (as indicated by determining the SG of the spent acid and its nitric acid content) agitation is stopped and the reaction mixture allowed to stand, when the nitrobenzene separates out as a top layer. The spent acid is dropped to a lead-lined 'egg' from which it is blown by compressed air to a storage tank. Spent acid is agitated with benzene in a nitrator (the extract is used in the next nitration). The nitrobenzene is blown to a washer where it is washed with water (if required for aniline manufacture) and further washed with sodium carbonate solution and vacuum distilled if it is to be sold. Alternatively, the separation of the spent acid from the nitrobenzene may be carried out in a separator. Nitrobenzene is also manufactured by continuous processes. Other important nitrations include that of toluene (see later), chlorobenzene and acetyl-p-toluidine, the nitration product of the latter yielding, on hydrolysis, the important pigment intermediate, 3-nitro-4-aminotoluene.

Nitrobenzene is toxic by absorption through the skin and in vapour form. In addition to its effect on the blood (it causes anaemia) it also affects the nervous system. Other nitro-compounds are also toxic to varying degrees. Another hazard which must be guarded against in nitration plants is nitrous fumes, which can cause serious lung damage.

Uses of nitrobenzene

Aniline. The greater part of the nitrobenzene produced is reduced to aniline (p 115)

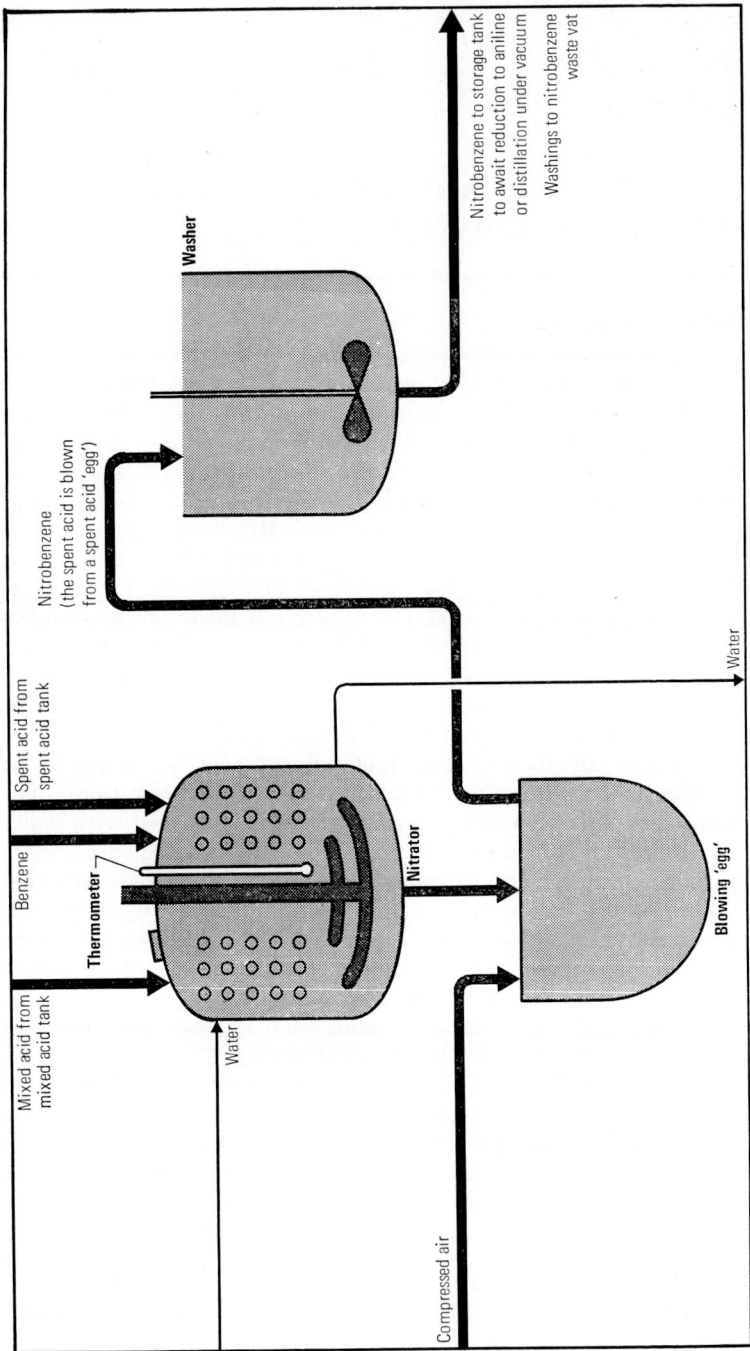

Mixed acid from mixed acid tank

Benzene

Spent acid from spent acid tank

Thermometer

Water

Nitrator

Nitrobenzene (the spent acid is blown from a spent acid 'egg')

Washer

Compressed air

Blowing 'egg'

Water

Nitrobenzene to storage tank to await reduction to aniline or distillation under vacuum

Washings to nitrobenzene waste vat

Other uses. These include the production of other dyestuff intermediates such as *m*-nitroaniline, metanilic acid (p 188) and black dyestuffs known as Nigrosines which are used in polishes

Nitrotoluenes

The preparation and uses of the principal nitrotoluenes are given in the scheme on p 114.

The mixture of dinitrotoluenes produced on p 114 (*i.e.* prior to crystallization from toluene) is used for the preparation of mixed 2,4- and 2,6-tolylenedi-isocyanates:

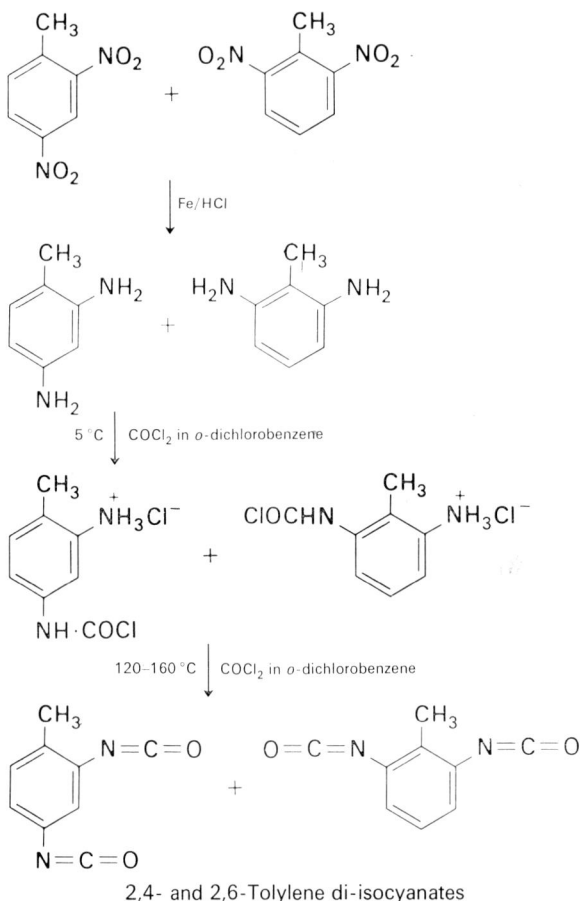

2,4- and 2,6-Tolylene di-isocyanates

The di-isocyanates on prolonged exposure in concentrations exceeding one part in ten million cause irritation of the eyes, throat

FIG. 23. (Facing page) Nitration of benzene—batch plant.

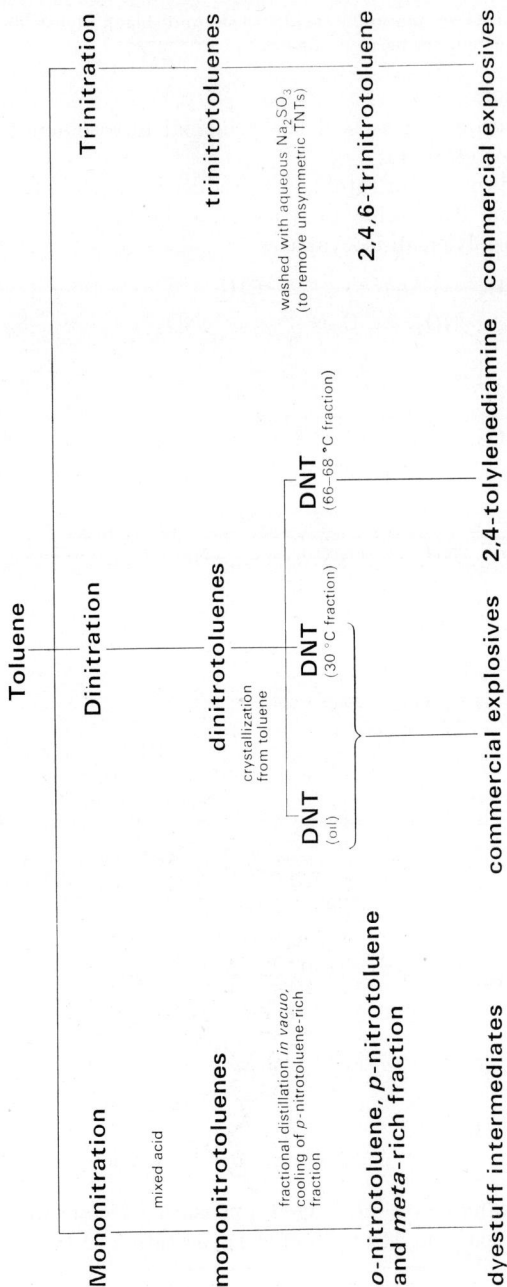

Toluene

Mononitration

mixed acid

mononitrotoluenes

fractional distillation *in vacuo*, cooling of *p*-nitrotoluene-rich fraction

o-nitrotoluene, *p*-nitrotoluene and *meta*-rich fraction

dyestuff intermediates

Dinitration

dinitrotoluenes

crystallization from toluene

DNT (oil)

DNT (30 °C fraction)

commercial explosives

DNT (66–68 °C fraction)

2,4-tolylenediamine

Trinitration

trinitrotoluenes

washed with aqueous Na_2SO_3 (to remove unsymmetric TNTs)

2,4,6-trinitrotoluene

commercial explosives

and lungs. They are used in the manufacture of flexible and rigid polyurethane foams. Production of non-rigid foams in the UK in 1968 was 34 800 t.

Aniline[3,4]

Aniline, $C_6H_5NH_2$, is mainly produced by the reduction of nitrobenzene, but it is also manufactured by the amination of phenol.

The reduction of nitrobenzene is effected (a) by means of iron borings and *dilute* hydrochloric acid (Béchamp process) or (b) by catalytic hydrogenation. The iron reduction process[3] has been in use for more than a century. It is carried out in a lagged iron

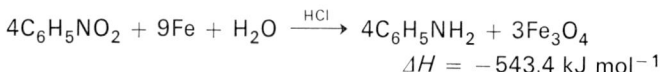

$$4C_6H_5NO_2 + 9Fe + H_2O \xrightarrow{\text{HCl}} 4C_6H_5NH_2 + 3Fe_3O_4$$
$$\Delta H = -543.4 \text{ kJ mol}^{-1}$$

vessel partly lined with brickwork, fitted with a heavy stirrer and connected to a reflux condenser. The process is controlled by adding the iron borings gradually. When reduction is complete the reaction mixture is neutralized by addition of lime and the aniline recovered by steam distillation. Alternatively after neutralization, the bulk of the aniline is siphoned off and live steam passed into the vessel to distil off the remaining aniline. The black oxide residue, after suitable treatment, can be sold to pigment manufacturers. The flow diagram for the process is shown in *Fig. 24*. Although the Béchamp process is becoming obsolescent for reducing nitrobenzene on account of its batch nature, it is still largely used for the production of other amino compounds (such as the toluidines and xylidines) which are required in smaller quantities than aniline. Catalytic hydrogenation of nitrobenzene is carried out in the vapour phase with the catalyst in the form of a fixed bed or a fluidized bed. A copper catalyst on a carrier such as silica gel is said to be suitable for fluid bed processes, and nickel sulphide on alumina for fixed bed processes. The reaction is highly exothermic and efficient removal of the heat of reaction is essential to prevent the reduction going all the way to benzene, with elimination of the nitrogen as ammonia. The effluent leaving the catalyst chamber is cooled to condense the aniline and the unused hydrogen is recycled. Aniline is now produced in the UK by catalytic hydrogenation of nitrobenzene but no details are available.

Aniline has also been manufactured by the amination of chlorobenzene: other dyestuff intermediates are manufactured by heating

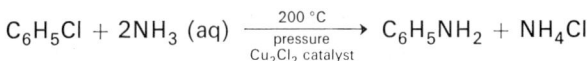

$$C_6H_5Cl + 2NH_3 \text{ (aq)} \xrightarrow[\substack{\text{pressure} \\ \text{Cu}_2\text{Cl}_2 \text{ catalyst}}]{200\,°C} C_6H_5NH_2 + NH_4Cl$$

halogenobenzenes with concentrated aqueous ammonia under pressure (no catalyst is used) *e.g. o-* and *p*-nitroanilines from *o-* and

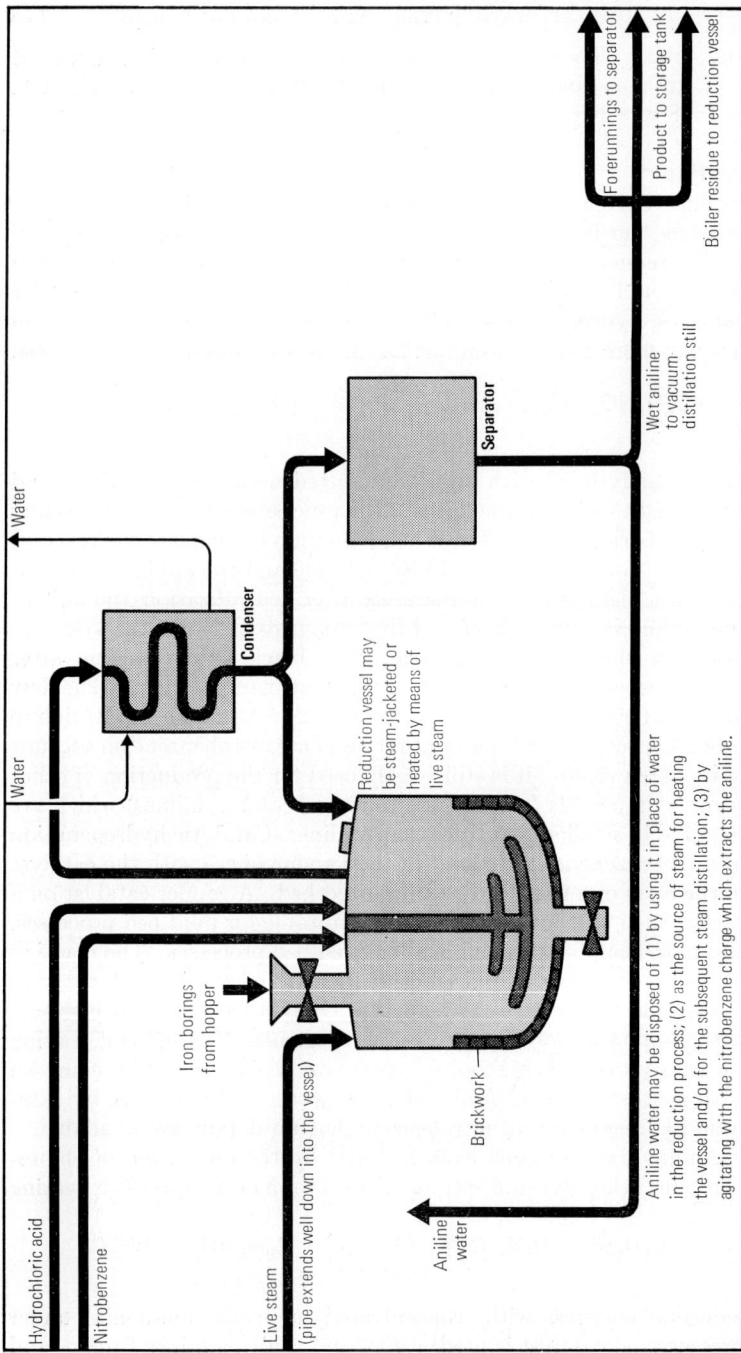

Hydrochloric acid

Nitrobenzene

Iron borings from hopper

Live steam
(pipe extends well down into the vessel)

Water

Water

Condenser

Reduction vessel may be steam-jacketed or heated by means of live steam

Brickwork

Aniline water

Separator

Wet aniline to vacuum distillation still

Forerunnings to separator

Product to storage tank

Boiler residue to reduction vessel

Aniline water may be disposed of (1) by using it in place of water in the reduction process; (2) as the source of steam for heating the vessel and/or for the subsequent steam distillation; (3) by agitating with the nitrobenzene charge which extracts the aniline.

p-nitrochlorobenzenes. Aniline is reported to be manufactured in Japan in a 20 000 t/a plant by the amination of phenol, the latter now being a sufficiently cheap raw material:

$$C_6H_5OH + NH_3 \longrightarrow C_6H_5NH_2 + H_2O$$

Details of the process are not available.

Aniline and other aromatic amines are very toxic by inhalation of the vapour and by absorption through the skin. It combines with the red colouring matter of the blood to give methaemoglobin which unlike oxyhaemoglobin is unable to give up oxygen to the tissues. Thus, the immediate result of poisoning is cyanosis (blue lips and finger tips), anaemia developing later.

Uses of aniline

Accelerators for the vulcanization of rubber and antioxidants for rubber (60 per cent)

Production of di-isocyanates (for rigid polyurethane foams)

Dyestuff intermediates and dyestuffs (12–14 per cent)

Other uses. These include the production of the photographic developer, hydroquinone, cyclohexylamine (manufactured in the UK by catalytic hydrogenation of aniline), sulphonamide drugs, the initiator explosive Tetryl and aniline–formaldehyde resins

References

1. E. Kilner and D. M. Samuel, *Applied organic chemistry*, p 249. London: Macdonald & Evans, 1960.
2. 'Nitration of benzene', *Unit processes in organic synthesis*, 5th edn, (P. H. Groggins ed.), p 107. New York: McGraw–Hill, 1958.
3. 'Manufacture of aniline', *Encyclopedia of chemical technology*, 3rd edn, (P. E. Kirk and D. F. Othmer eds), vol 12, p 419. New York: Interscience, 1963.
4. A. H. Jubb and M. C. McCarthy, 'Manufacture of aniline', *Educ. Chem.*, 1970, **7**, 113.

FIG. 24. (Facing page) Reduction of nitrobenzene with iron and dilute hydrochloric acid.

12. Sulphonic Acids

Sulphonation of benzene to produce benzenesulphonic acid was formerly an important process; the acid in the form of its sodium salt was used for the manufacture of phenol by caustic fusion. This latter process is no longer economic and is now obsolete in the UK. However, the sulphonation of other aromatic compounds is still important, *e.g.*

(1) The disulphonation of benzene:

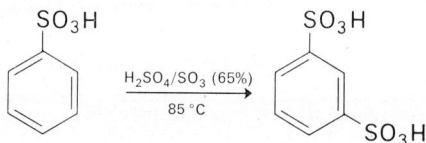

The disodium salt of the product is used for the production of resorcinol by caustic fusion.

(2) The sulphonation of nitrobenzene:

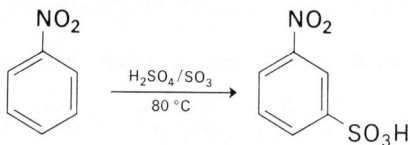

Reduction of the product gives metanilic acid, used as a dyestuff intermediate.

(3) The sulphonation of naphthalene, 2-naphthol and anthraquinone.

The β-sulphonation of naphthalene is effected at 160 °C with a 20 per cent excess of concentrated sulphuric acid. The 15 per cent of α-acid which is also formed is removed by passing live steam into the hot reaction mixture, which hydrolyses the α-acid to naphthalene (steam distilled out).

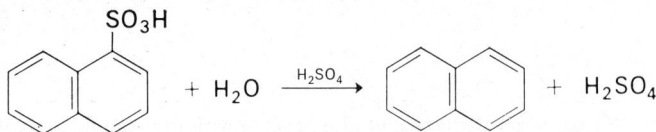

The β-acid is isolated as its sodium salt by liming-out (*see* below). Fusion of the sodium salt with caustic soda yields the sodium salt of β-naphthol, the latter being produced in greater quantity than any other naphthalene derivative.[1]

The sulphonation of β-naphthol and anthraquinone under varying conditions leads to a number of important dyestuff intermediates.

(4) A very important process is the sulphonation of long-chain alkylbenzenes to produce anionic surface-active agents:

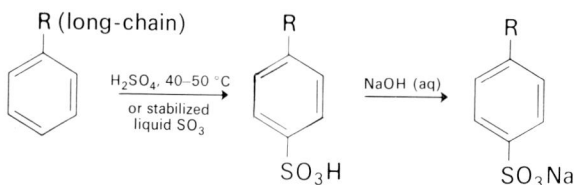

R (long-chain) $\quad\xrightarrow[\substack{\text{or stabilized}\\\text{liquid } SO_3}]{H_2SO_4,\ 40\text{--}50\ ^\circ C}\quad$ R—SO_3H $\quad\xrightarrow{\text{NaOH (aq)}}\quad$ R—SO_3Na

Aromatic compounds are still most commonly sulphonated with sulphuric acid or oleum, but stabilized liquid sulphur dioxide is finding increasing use as a sulphonating agent, particularly for the sulphonation of long-chain alkylbenzenes.

Sulphonic acids are isolated by *liming out*,

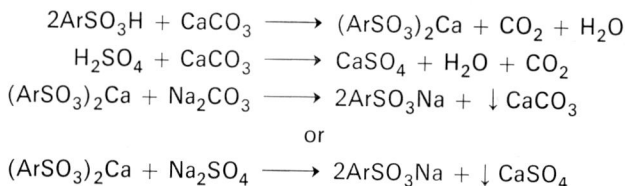

$$2ArSO_3H + CaCO_3 \longrightarrow (ArSO_3)_2Ca + CO_2 + H_2O$$
$$H_2SO_4 + CaCO_3 \longrightarrow CaSO_4 + H_2O + CO_2$$
$$(ArSO_3)_2Ca + Na_2CO_3 \longrightarrow 2ArSO_3Na + {\downarrow} CaCO_3$$

or

$$(ArSO_3)_2Ca + Na_2SO_4 \longrightarrow 2ArSO_3Na + {\downarrow} CaSO_4$$

and by *salting out*:[2]

$$ArSO_3H + NaCl\ (aq) \rightleftharpoons {\downarrow} ArSO_3Na + HCl$$

References

1. E. Kilner and D. M. Samuel, *Applied organic chemistry*, p 283. London: Macdonald and Evans, 1960.
2. Ref. 1, p 298.

13. Phenols

Phenol[1,2]

The amount of phenol available from carbonization of coal is limited by the demand for gas and coke, and has for a long time been insufficient to meet the demand. Production of phenol from coal in the UK in 1967 amounted to 17 300 t and by 1975 it is expected that only about 8000 t/a will be available from this source. The UK production of synthetic phenol in 1966 was 77 940 t (the production for later years is not available). Current UK consumption of phenol amounts to about 100 000 t/a and by 1975 it is estimated that the demand will rise to 140 000 t/a.

The oldest synthetic process for the production of phenol involved sulphonation of benzene, neutralization of the sulphonation mixture with sodium sulphite, fusion of the sodium sulphonate with caustic soda, and decomposition of the phenoxide by sulphur dioxide

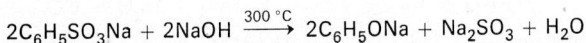

$$2C_6H_5SO_3Na + 2NaOH \xrightarrow{300\ °C} 2C_6H_5ONa + Na_2SO_3 + H_2O$$

(produced by the neutralization of the sulphonation mixture with sodium sulphite). The process depended on a ready sale for the by-products sodium sulphite and sulphur dioxide. It was operated in the UK until a few years ago. The chlorination process is a three-step process involving the chlorination of benzene, hydrolysis of the chlorobenzene, and decomposition of the phenoxide by hydrochloric

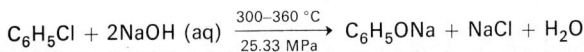

$$C_6H_5Cl + 2NaOH\ (aq) \xrightarrow[25.33\ MPa]{300-360\ °C} C_6H_5ONa + NaCl + H_2O$$

acid. The process was last operated in the UK in 1964. Phenol is also made by a two-stage process from benzene:

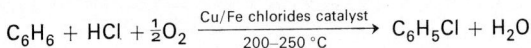

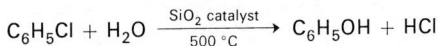

$$C_6H_6 + HCl + \tfrac{1}{2}O_2 \xrightarrow[200-250\ °C]{Cu/Fe\ chlorides\ catalyst} C_6H_5Cl + H_2O$$

$$C_6H_5Cl + H_2O \xrightarrow[500\ °C]{SiO_2\ catalyst} C_6H_5OH + HCl$$

The hydrochloric acid produced in the second stage is re-used in the first stage. This process (Raschig process) has never been operated in Britain. It is believed that a modification of the process (Hooker process) is still operated in the US.

The cumene process, more economic than any of the above processes, now accounts for all the synthetic phenol produced in the UK. It is represented by the following equations:

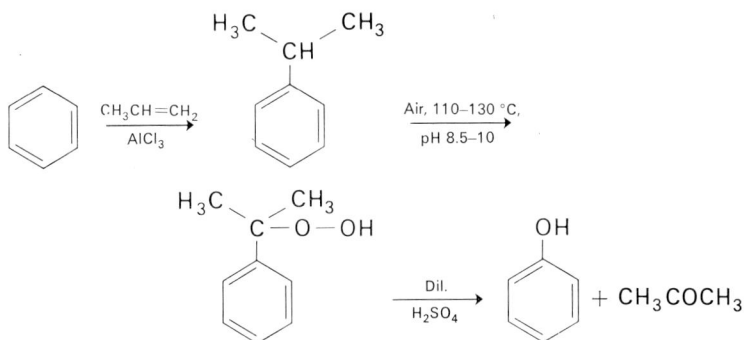

The decomposition of the hydroperoxide probably takes place by the following mechanism:

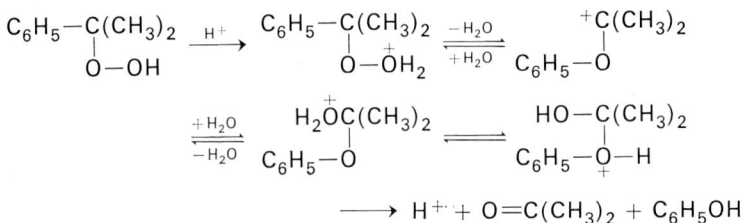

Acetophenone and phenyldimethylcarbinol are by-products of the oxidation, the latter compound undergoing dehydration during the fission process to give α-methylstyrene. The latter is available in commercial quantities from the cumene process, one of its main outlets being in the production of co-polymers, *e.g.* as a replacement for styrene in ABS polymers giving a product with a higher softening point; it is also used as a partial replacement for styrene in the production of glass-reinforced polyester resins, giving a cross-linked product with improved properties. The economics of the cumene process are also helped if acetone can be sold at a reasonable price. At present the demand for acetone seems to be keeping pace with that for phenol.

Figure 25 is the flow diagram of the BP Chemicals International Ltd phenol process. Cumene is oxidized with air under carefully controlled conditions. The oxidate is concentrated to about 80 per cent cumene hydroperoxide in special equipment (the distillate is recycled to the oxidizers) and is then treated with an acid catalyst in a cleavage reactor of special design, combining high efficiency with safety. After removal of the acid catalyst the reaction mixture is fractionated to produce acetone and phenol products. Special distillation

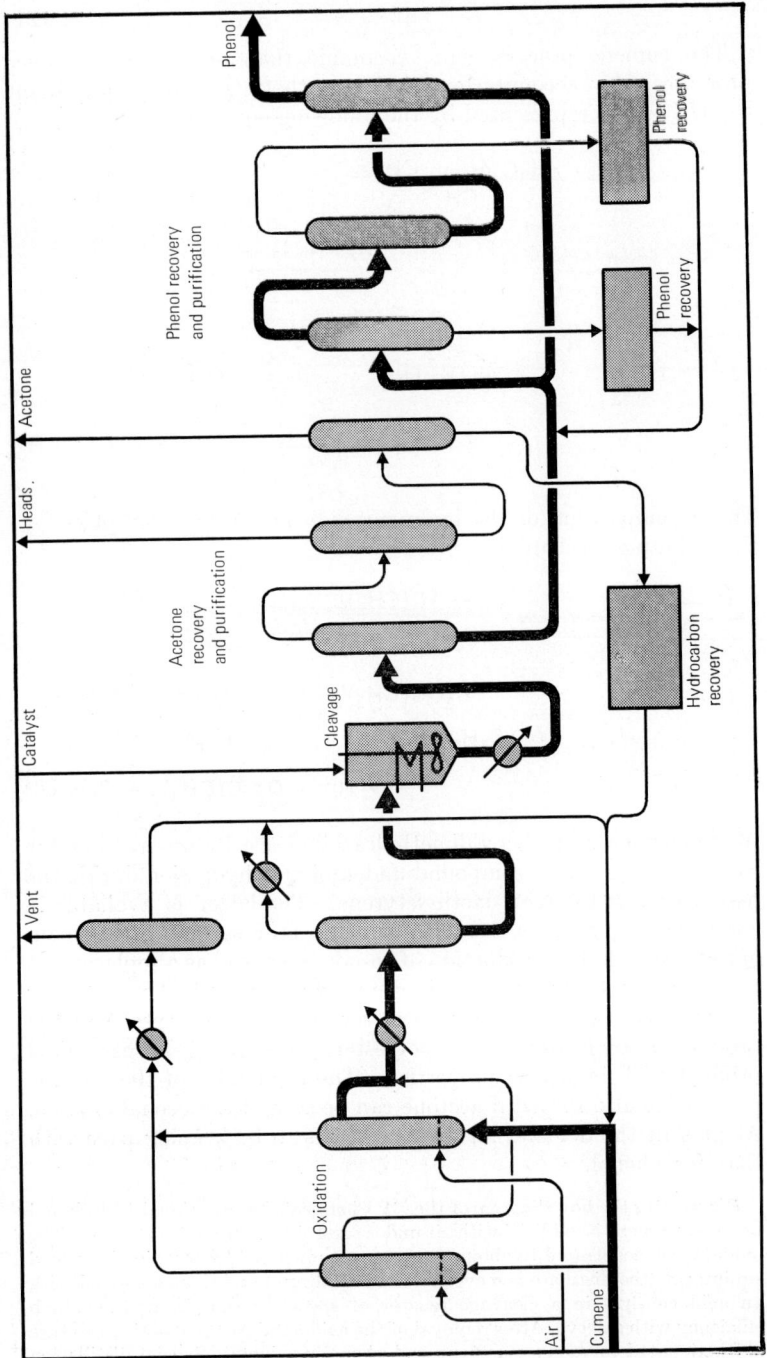

Phenol

Phenol recovery
and purification

Phenol recovery

Phenol recovery

Acetone

Heads

Acetone recovery
and purification

Catalyst

Cleavage

Hydrocarbon recovery

Vent

Oxidation

Air

Cumene

units are employed to remove hydrocarbons and other impurities to very low levels. The high-boiling residues can be distilled to recover residual free phenol and phenol liberated by cracking of complex hydrocarbons. The α-methyl-styrene by-product can be hydrogenated to cumene, re-cycled to the oxidizers or isolated. The overall conversion efficiency of cumene to phenol can be as high as 92.5 per cent. Approximately two-thirds (on a weight basis) as much acetone as phenol is produced.

Phenol has also been manufactured in the US since 1960 by the following process (Dow process):

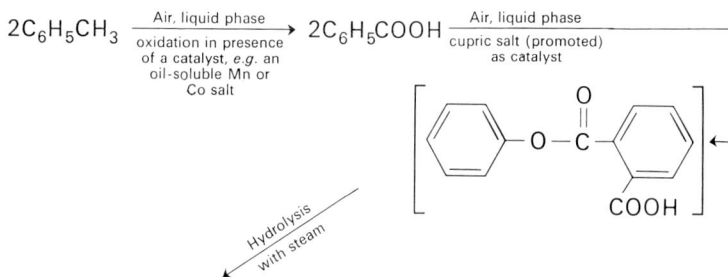

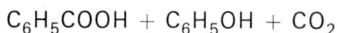

$$C_6H_5COOH + C_6H_5OH + CO_2$$

The process can also be used to produce m-cresol from p-xylene, and o- and p-cresols from m-toluic acid.

Another route to phenol is as follows:

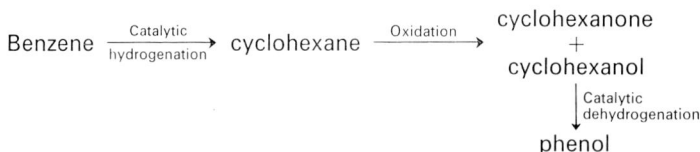

It is worth mentioning that all major companies have tried the direct oxidation of benzene to phenol, but so far unsuccessfully.

Uses of phenol

Phenolic resins (from phenols and formaldehyde). In addition to phenol, m- and p-cresol mixtures are also used in large quantities for the production of phenolic resins. The production of phenolic resins (72 300 t in the UK in 1969) probably accounts for about 60 per cent of the phenol produced

Alkyl phenols. These include p-t-butylphenol and p-octylphenol, which are produced by alkylating phenol with isobutene and di-isobutene, respectively. Both are used for the production of oil-soluble phenolic resins, p-octylphenol also being used for the production of non-ionic surface-active agents

FIG. 25. (Facing page) Phenol production—BP Chemicals International Ltd. (Reproduced from the November 1969 issue of *Hydrocarbon Processing*, p 214, by permission of the Gulf Publishing Co., Houston, Texas.)

Chlorophenols. The chlorination of phenol can be controlled by carrying out the process in the absence of water. According to the quantity of chlorine passed into the molten phenol it is possible to obtain monochlorophenols, 2,4-dichlorophenol or 2,4,6-trichlorophenol as chief product, whilst prolonged chlorination in the presence of anhydrous ferric chloride as catalyst yields pentachlorophenol. 2,4-Dichlorophenol is used in large quantities for making derivatives of 2,4-dichlorophenoxyacetic acid, which are used as selective weed killers. Trichlorophenol is well-known as an antiseptic, whilst the sodium salt of pentachlorophenol is important as a timber preservative

Caprolactam (for nylon 6). The greater part of the world production of caprolactam is based on cyclohexanone derived from benzene as starting material, but phenol is still used as a starting material, being hydrogenated to cyclohexanol which on oxidation or dehydrogenation yields cyclohexanone

Bisphenol A (2,2-bis-p-hydroxyphenylpropane).[3] This is made by condensing phenol with acetone in the presence of hydrogen chloride as catalyst:

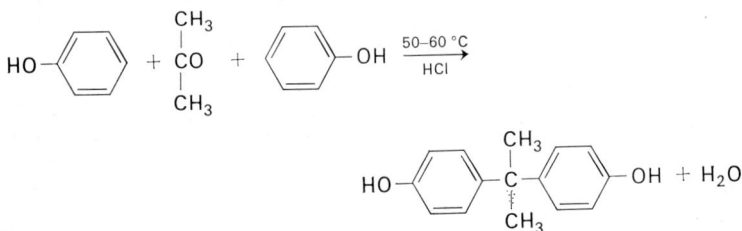

$$HO-\langle\!\!\!\bigcirc\!\!\!\rangle + \underset{\underset{CH_3}{|}}{\overset{\overset{CH_3}{|}}{CO}} + \langle\!\!\!\bigcirc\!\!\!\rangle-OH \xrightarrow[HCl]{50\text{--}60\,^\circ C}$$

$$HO-\langle\!\!\!\bigcirc\!\!\!\rangle-\underset{\underset{CH_3}{|}}{\overset{\overset{CH_3}{|}}{C}}-\langle\!\!\!\bigcirc\!\!\!\rangle-OH + H_2O$$

At present the UK output of bisphenol A is 20 000 t/a but demand is increasing rapidly. It is used in the production of epoxy resins (p 25) and polycarbonate plastics. The latter, manufactured from bisphenol A and phosgene, are used for electrical insulation and as a replacement for metals (they have very high impact strengths)

Miscellaneous. These include the production of salicylic acid (for aspirin and dyestuffs), cyclohexanol, photographic chemicals and triphenyl phosphate

Phenol can be toxic by absorption, in solution, through the skin (which takes place rapidly) or by inhalation of the vapour. Its caustic effect upon the skin is well known.

Cresols and xylenols

Until comparatively recently cresols and xylenols were obtained almost exclusively as by-products of coal-carbonization processes in the UK. The most important source has been coal-tar derived from gas-works carbonization, which will not be available after 1975. Current production of cresols in the UK is about 70 000 t/a, by 1972 production will probably have fallen to less than 55 000– 60 000 t/a, and by 1975 probably to 50 000 t/a. There is at present a shortage of cresols and this has stimulated increased production from coal by low-temperature carbonization and by synthetic methods. One firm is erecting an o-cresol plant of 5000 t/a capacity which is expected to come into production by 1971. Details of the processes used in the UK are not available. The Koppers *o*-cresol process consists in *C*-methylating phenol over a catalyst at elevated temperature:

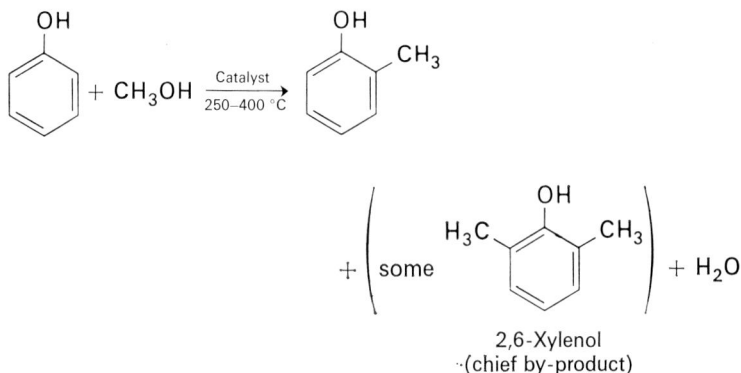

$$\text{phenol (OH)} + CH_3OH \xrightarrow[\text{250–400 °C}]{\text{Catalyst}} \text{o-cresol (OH, CH}_3\text{)}$$

$$+ \text{some} \left(\text{2,6-Xylenol: OH, } H_3C \text{ and } CH_3 \right) + H_2O$$

2,6-Xylenol
(chief by-product)

In the US *p*-cresol is synthesized from *p*-cymene, $CH_3C_6H_4p$-CH$(CH_3)_2$, by a process analogous to that used for the production of phenol from cumene. For some of the applications of cresols, such as the production of plasticizers, synthetic isopropylphenol has become a competitor.

Uses of cresols and cresylic acid

Phenolic resins (see uses of phenol). These probably account for about 50 per cent of the *m*- and *p*-cresol production

Tricresyl phosphate, $(CH_3C_6H_4O)_3PO$. This is produced by heating a *m*- and *p*-cresol mixture with phosphorus oxychloride. It is used as a plasticizer for polyvinyl chloride, cellulose acetate and nitrocellulose. In addition to its plasticizing effect it reduces the flammability of these substances. UK production of phosphates (aliphatic and aromatic) in 1970 amounted to 27 950 t

Octylcresol. This is produced by alkylating cresylic acid with di-isobutene. It is important for making non-ionic surface-active agents by condensation with ethylene oxide

Miscellaneous uses of cresylic acid. These include ore flotation, metal cleaning and production of disinfectants and methylcyclohexanols

o-Cresol is used for the production of the weed killers dinitro-*o*-cresol and 2-methyl-4-chlorophenoxyacetic acid (MCPA), the latter being a selective weed killer. It is also used for making *o*-cresotinic acid, *i.e.* 2-hydroxy-3-methylbenzoic acid (dyestuff intermediate)

m-Cresol is used for making germicides and fine chemicals

p-Cresol is used for making antioxidants for motor spirit and polyolefins, and perfume ingredients

The remarks about the toxic properties of phenol apply to the cresols and xylenols which also have a caustic effect upon the skin.

References

1. R. B. Stobaugh, 'Phenol; how, where, who—future', *Hydrocarb. Process.*, 1966, **45**(1), 143.
2. A. S. Banciu, 'Phenol manufacture', *Chem. Process. Engng*, 1967, **48**(1), 31.
3. Hooker Chemical Co, 'Bisphenol A', *Hydrocarb. Process.*, 1967, **46**(11), 152.

14. Aromatic Alcohols, Aldehydes, Ketones and Acids

In terms of quantity produced by far the most important compounds discussed in this section are the aromatic acids and their derivatives. Aromatic alcohols, aldehydes and ketones are produced in much lesser quantities since they have no really large scale outlets.

Benzyl alcohol[1]

Benzyl alcohol, $C_6H_5CH_2OH$, is produced by chlorinating toluene to benzyl chloride and hydrolysing the latter. The side-chain chlorination of toluene is carried out by passing chlorine into boiling toluene (the reaction is exothermic); ultraviolet light catalyzes the process. In order to minimize the formation of benzylidene chloride, the process is not allowed to go to completion and is controlled by density measurements. The crude product is fractionally distilled to separate the benzyl chloride from unchanged toluene and benzylidene chloride. The hydrolysis of benzyl chloride is effected by heating with aqueous sodium carbonate solution under reflux conditions:

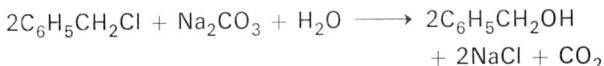

$$2C_6H_5CH_2Cl + Na_2CO_3 + H_2O \longrightarrow 2C_6H_5CH_2OH + 2NaCl + CO_2$$

The crude product is purified by fractional distillation under reduced pressure. Benzyl alcohol is used chiefly for the production of its esters which are used in flavouring essences and in perfumery. It finds limited use as a high-boiling solvent.

Benzaldehyde[2]

Benzaldehyde, C_6H_5CHO, is manufactured by the hydrolysis of benzylidene chloride with an equivalent of sodium carbonate in

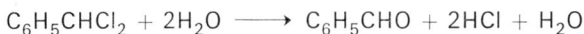

$$C_6H_5CHCl_2 + 2H_2O \longrightarrow C_6H_5CHO + 2HCl + H_2O$$

aqueous solution or with 70 per cent sulphuric acid, and also by the vapour-phase and liquid-phase oxidation of toluene. The chlorination route is obviously still important since a UK firm is to erect a large scale plant for the production of benzyl alcohol and benzaldehyde by this route. The advantage of the oxidation processes is that they yield a product free from nuclear chlorine as required in flavours and perfumery. Benzaldehyde is chiefly used as a flavouring material ('almond essence') and for making dyes (*e.g.* Malachite Green, Brilliant Green and Disulphine Blue) and perfumery ingredients.

Acetophenone

Acetophenone, $C_6H_5COCH_3$, was formerly manufactured by the Friedel–Crafts process from benzene and acetic anhydride in the presence of aluminium chloride as catalyst, but is now manufactured by the catalytic air-oxidation of ethylbenzene.[3] Acetophenone is also formed as a by-product (3 per cent) in the manufacture of phenol from cumene hydroperoxide, and this source is likely to become more important as the quantity of phenol manufactured increases. It is mainly used in perfumery and to a small extent as an intermediate and solvent.

Benzoic acid

Benzoic acid, C_6H_5COOH, has been manufactured by the catalytic decarboxylation of phthalic anhydride at elevated temperature and by the hydrolysis of benzotrichloride produced by the chlorination of toluene. The acid is made more economically now by the liquid phase catalytic air oxidation of toluene at elevated temperature under pressure:[4]

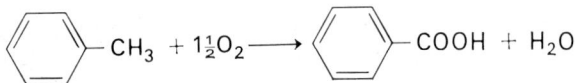

$$\langle\!\!\rangle\!\!-CH_3 + 1\tfrac{1}{2}O_2 \longrightarrow \langle\!\!\rangle\!\!-COOH + H_2O$$

The process has been operated in the UK since about 1966 in a plant of 6000 t/a capacity.

The flow diagram for the Snia Viscosa process for the manufacture of benzoic acid is shown in *Fig. 26.* Toluene and air are fed to the reactor in which oxidation takes place at 150–170 °C and at 1.013 MPa pressure in the presence of an aqueous solution of cobalt acetate catalyst (the concentration of catalyst is kept below 100 ppm). The toluene conversion is maintained at 30–35 per cent. The liquid from the reactor passes to a two column distillation system operated at atmospheric pressure. Unchanged toluene is taken off from the head of the first column and benzyl alcohol and benzaldehyde by-products from the head of the second column. The toluene and by-products are recycled to the reactor. Benzoic acid is taken off as a side-stream from the second column at a purity greater than 99 per cent.

Benzoic acid is used largely in the form of its derivatives, the sodium salt accounting for about 80 per cent of the production. This is widely used as a preservative, in very small quantities (120–600 ppm), in food products (particularly fruit squashes) and in pharmaceuticals. It is also used as a corrosion

Fig. 26. (Facing page) Benzoic acid production—Snia Viscosa process. (Reproduced from the November 1969 issue of *Hydrocarbon Processing*, p 156, by permission of the Gulf Publishing Co., Houston Texas.)

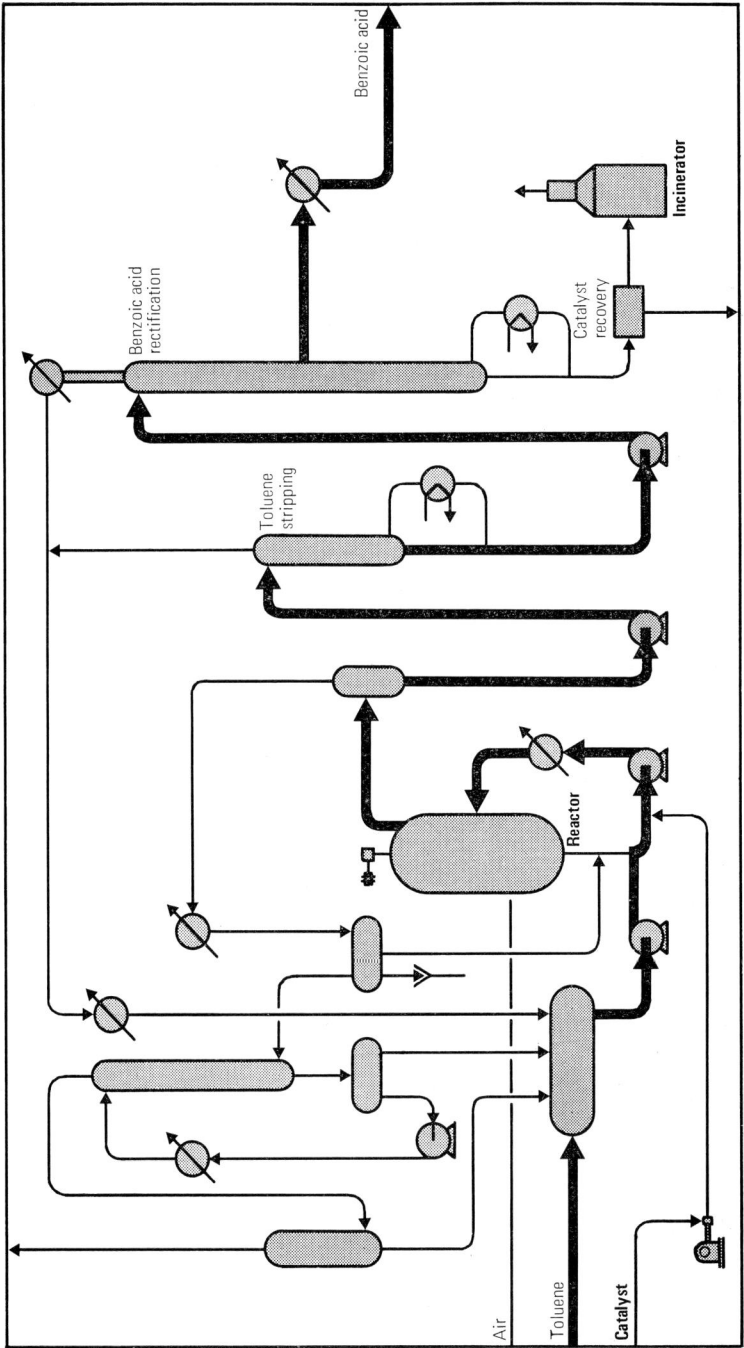

inhibitor, *e.g.* for impregnating paper to contain needles, razor blades and so forth, in industrial processes, and in conjunction with sodium nitrite to prevent corrosion by glycol in antifreeze solutions. Benzoic acid is also used for the manufacture of dyes, esters, plasticizers and 3,5-dinitrobenzoyl chloride (used in the purification of vitamin D_2).

Phthalic anhydride[5-7]

Phthalic anhydride is manufactured by the air oxidation of naphthalene or *o*-xylene over a bed of vanadium pentoxide catalyst on silica or alumina at elevated temperatures:

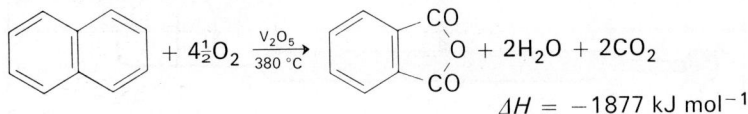

$$\Delta H = -1877 \text{ kJ mol}^{-1}$$

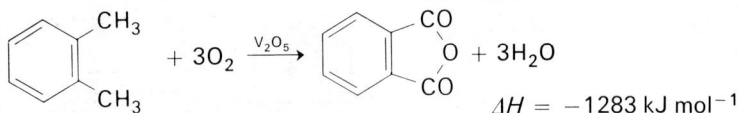

$$\Delta H = -1283 \text{ kJ mol}^{-1}$$

In the case of naphthalene oxidation the catalyst is employed in the form of a fixed or fluidized bed but *o*-xylene oxidation has so far only been effected over a fixed-bed catalyst. It is believed that oxidation of naphthalene takes place as follows:

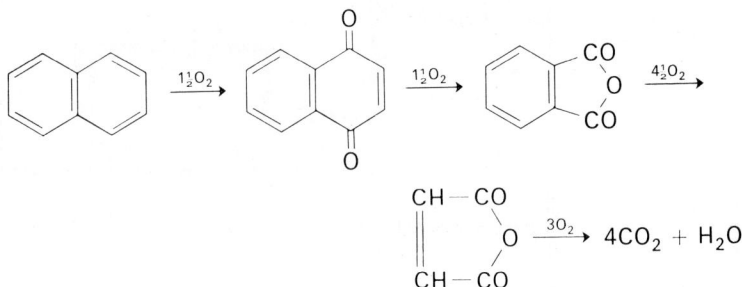

In practice, excess air is used to cut down the chance of an explosion. The oxidation is highly exothermic and a short contact time and an adequate means of removing the heat of reaction (involving specially constructed reaction chambers) are essential to minimize over-oxidation of naphthalene, with production of maleic anhydride and oxidation products of the latter. Prior to 1964 all the phthalic anhydride produced in the UK was made from naphthalene. About 75 per cent of the naphthalene production is used for making phthalic anhydride.

FIG. 27. (Facing page) Production of phthalic anhydride from naphthalene—Sherwin–Williams/Badger, The Badger Co. Inc. (Reproduced from the November 1969 issue of *Hydrocarbon Processing* p 217, by permission of the Gulf Publishing Co., Houston, Texas.)

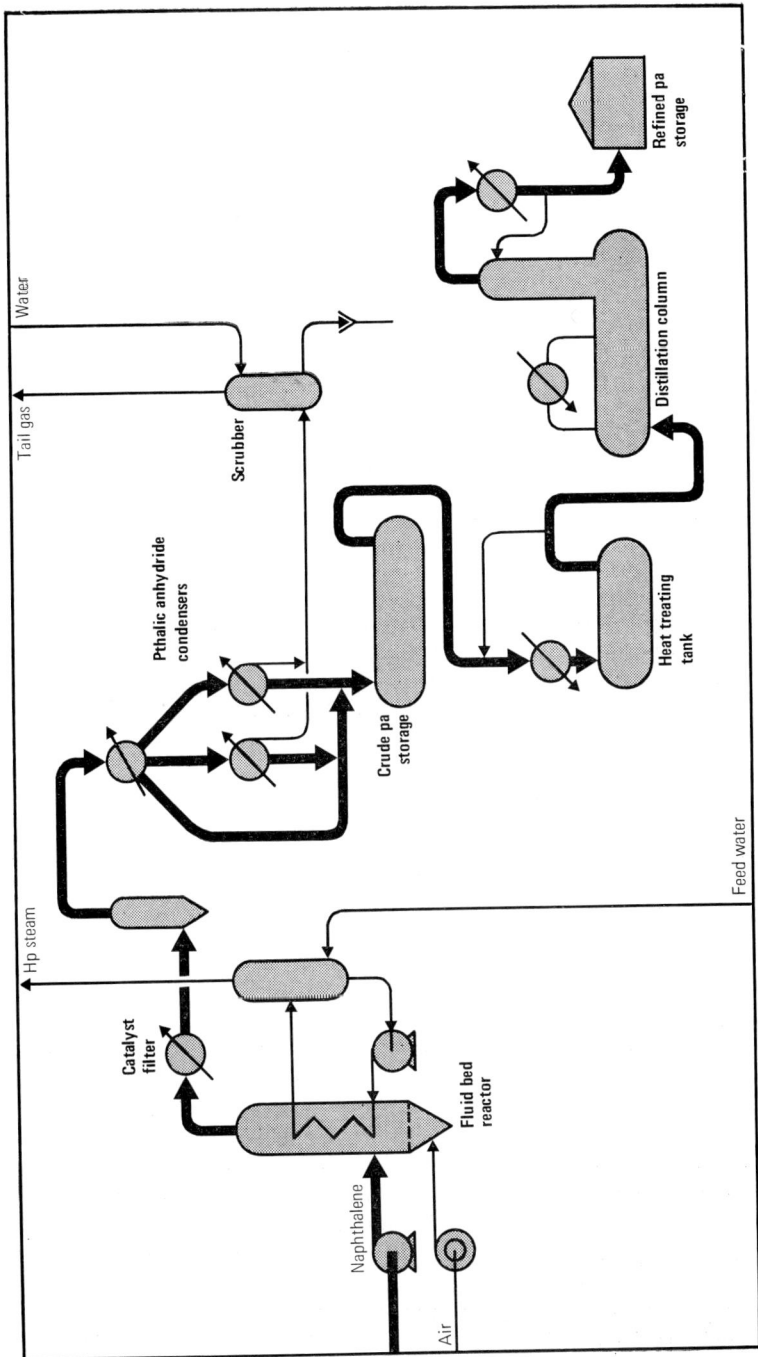

The flow-diagram of the Sherwin–Williams/Badger process for the oxidation of naphthalene is shown in *Fig. 27*. Liquid naphthalene is introduced into the base of the reactor where it is vaporized and distributed throughout the bed of catalyst by contact with the hot catalyst and the reactor air. A high degree of agitation and mixing takes place within the fluid bed thus ensuring the maintenance of a uniform temperature throughout the bed. The temperature in the reactor is controlled in the range 343–384 °C. The heat of reaction is removed by cooling tubes in the hot bed and high pressure steam is generated. The effluent from the reactor is passed through ceramic filter elements to remove entrained catalyst. The product is condensed as a liquid and a solid. In fluidized bed processes the air rate and the air-to-feed ratio are lower than in the case of fixed bed processes, with the result that 40–60 per cent of the crude product can be condensed directly as liquid product with consequent reduction of the load on the solid condensers and a proportionate decrease in plant costs as compared with fixed bed units. Fluidized bed processes are also better suited for very large scale production. The crude product is purified by a simple heat treatment and vacuum distillation. Yields are of the order of 98 lb of anhydride from 100 lb of naphthalene.

Since 1964 two o-xylene oxidation plants have come into operation in the UK. Naphthalene has been in short supply over the last few years whereas o-xylene has become available in increasing quantities. o-Xylene is cheaper than naphthalene and the fixed bed oxidation process is claimed to give a yield of anhydride equivalent to that obtained from naphthalene. On the other hand, the oxidation of o-xylene takes place under more drastic conditions than that of naphthalene. Thus, a relatively greater quantity is converted to carbon dioxide and water than is the case with naphthalene so that relatively more heat is liberated.

Production of phthalic anhydride in the UK in 1968 was 87 940 t.

Uses of phthalic anhydride

Dialkyl phthalates. These accounted for more than 50 per cent of the consumption of phthalic anhydride in the UK in 1968. Production of phthalates in the UK in 1970 amounted to 98 550 t. The most important are di-alphyl phthalates (from a mixture of C_7 and C_9 alcohols) and di-iso-octyl phthalate, production of which together accounts for about 75 per cent of the total production of phthalates. Other phthalates manufactured include dimethyl, diethyl, di-n-butyl, di-octyl and di-nonyl phthalates. The long-chain phthalates are used in large quantities for plasticizing polyvinyl chloride. The diethyl and di-butyl esters are important as plasticizers for cellulose acetate, the latter ester being also used for plasticizing polyvinyl acetate

Alkyd resins and polyester resins. Alkyd resins have already been referred to (p 25). Quite large quantities of anhydride are used for making these resins, but they are receiving increasing competition from thermosetting acrylic resins for the preparation of stoving enamels. The market for glass-reinforced polyester resins is expected to grow during the next few years

Other products. Phthalic anhydride is used for the production of anthraquinone dyes, phthalocyanine pigments and dyes, phenolphthalein and fluorescein. In some countries potassium phthalate is used for the manufacture of terephthalic acid by an isomerization process (see later)

Inhalation of phthalic anhydride dust can irritate the lungs.

Terephthalic acid[8]

Terephthalic acid is manufactured by the oxidation of p-xylene. The oxidation is effected in the liquid phase at elevated temperature with air under sufficient pressure (about 1.35 MPa) to maintain the p-xylene in the liquid phase in the presence of a catalyst, such as promoted cobalt acetate.

In the Raecke process use is made of isomerization:

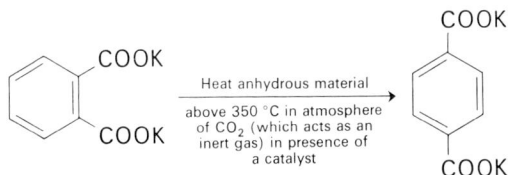

Heat anhydrous material above 350 °C in atmosphere of CO_2 (which acts as an inert gas) in presence of a catalyst

Potassium isophthalate may also be used whilst potassium benzoate yields potassium terephthalate and benzene.

Terephthalic acid is manufactured in very large quantities. It is almost exclusively used in the form of its dimethyl ester for the production of the polyester fibre, polyethylene terephthalate (Terylene). Polyester fibres now account for about 31 per cent of the production of synthetic fibres.

References

1. 'Benzyl alcohol', *Encyclopedia of chemical technology*, 3rd edn, (P. E. Kirk and D. F. Othmer eds), vol 3, p 440. New York: Interscience, 1964.
2. Ref. 1, vol 3 (1964), p 301.
3. British intelligence objectives sub-committee (BIOS), No. 1053. (A report on the chemical industry in Germany.)
4. *Eur. Chem. News*, 1963, **3**(73), 27.
5. R. Loudon and H. Harper, 'Phthalic anhydride', *Chemy Ind.*, 1961, 1143.
6. H. L. Riley, 'Phthalic anhydride', *Chemy Ind.*, 1966, 979.
7. R. F. Schwab and W. H. Doyle, 'Hazards in phthalic anhydride plants', *Chem. Engng Progress*, 1970, **66**(9), 49.
8. 'Recent advances in commercial terephthalic acid synthesis', *Chem. Age*, 1960, **84**(2140), 106.

15. Other Organic Compounds

Cyclohexanone[1]

Cyclohexanone is one of the most important alicyclic compounds produced on the heavy chemical scale. It is manufactured by the oxidation of cyclohexane, which in turn is obtained by the catalytic hydrogenation of benzene. In the Dutch State Mines 'Oxanon' process (*Fig. 28*) cyclohexane is oxidized with air in the presence of a catalyst at 150–160 °C and 0.8104–0.9117 MPa pressure to yield a mixture of cyclohexanol and cyclohexanone together with acids, esters and gaseous oxides of carbon.

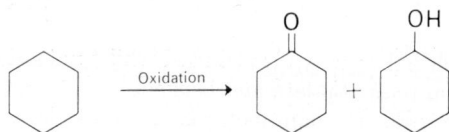

The reaction mixture is washed with sodium hydroxide solution, which dissolves the acids and saponifies the esters, and is then distilled to recover unchanged cyclohexane, which is recycled.

A modification of the oxidation step has been developed by the Institut Francais du Petrole, the process being carried out in the presence of boric acid. This forms a complex with cyclohexanol thus inhibiting further oxidation of the latter and improving the yield and conversion per pass. Hydrolysis of the complex liberates the cyclohexanol, the boric acid passing into the aqueous phase from which it is recovered.

Cyclohexanone and cyclohexanol are separated from the crude reaction product by distillation. The cyclohexanol is subjected to vapour phase dehydrogenation to cyclohexanone, which is returned to the distillation stage.

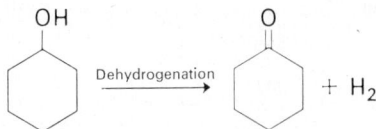

Fig. 28. (Facing page) Cyclohexanone from cyclohexane by Dutch State Mines 'Oxanon' process. (By courtesy of Simon–Carves Chemical Engineering Ltd.)

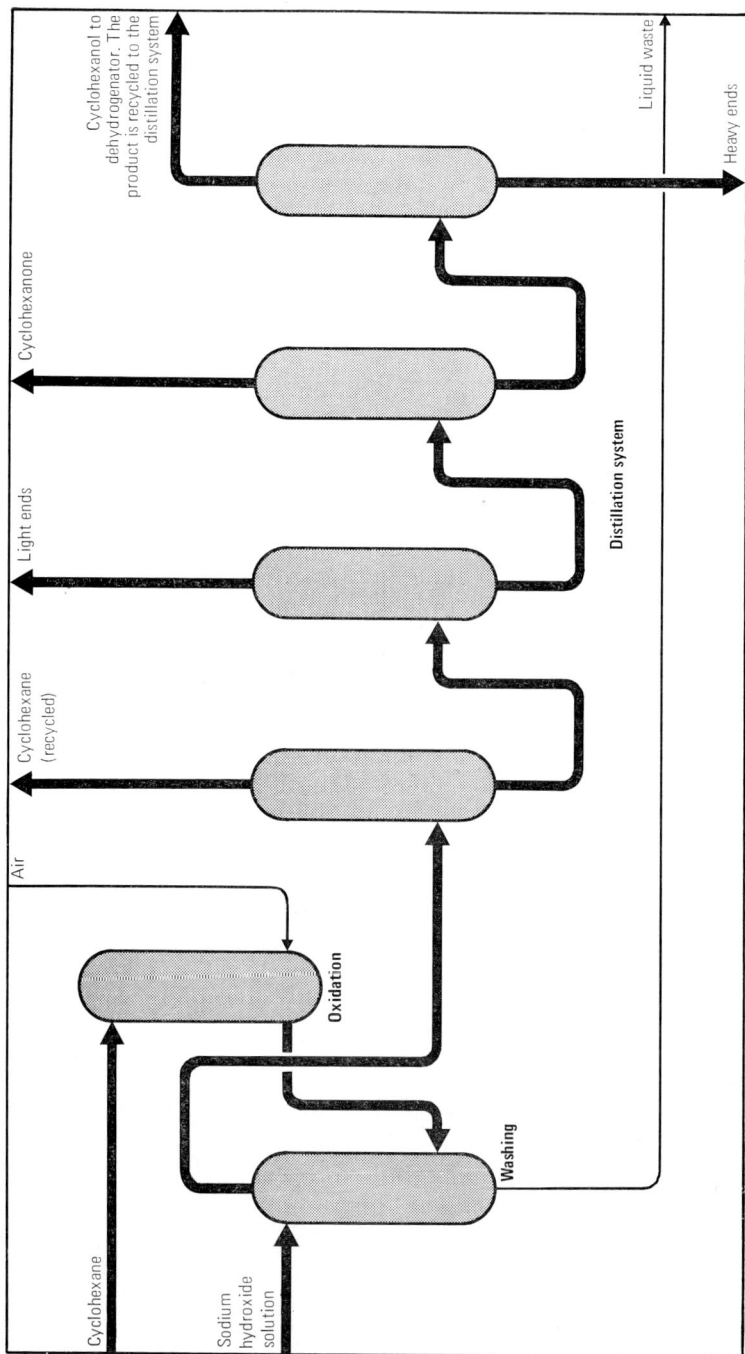

Air

Cyclohexane

Sodium hydroxide solution

Oxidation

Washing

Cyclohexane (recycled)

Light ends

Cyclohexanone

Cyclohexanol to dehydrogenator. The product is recycled to the distillation system

Liquid waste

Heavy ends

Distillation system

Cyclohexanone is produced in large quantities for making caprolactam, which is polymerized to polycaprolactam or nylon 6. The latter is valuable as a fibre and as a plastic material.

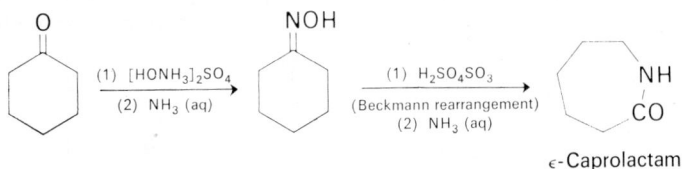

ε-Caprolactam

Dyestuffs and pigments

Synthetic dyestuffs and organic pigments are manufactured from the aromatic hydrocarbons benzene, toluene, xylenes and naphthalene, and from anthraquinone, the latter being produced in the UK by the oxidation of anthracene. In this monograph it has only been possible to deal with a small number of the intermediates (those produced in appreciable quantities) used for making dyestuffs and pigments.

Production of synthetic dyestuffs in the UK in 1969 amounted to 37 660 t and of pigment dyestuffs in 1968 to 10 000 t. The dyestuffs supplies in the UK contributed £20m. towards the UK exports in 1969.

Fig. 29. Britain's first caprolactam plant built by Simon–Carves Chemical Engineering Ltd for Nypro (UK) Ltd at Flexborough. (By courtesy of Simon–Carves Chemical Engineering Ltd.)

The production of intermediates for the dyestuffs and allied industries has been reviewed.[2-4] Recent developments in dyestuff chemistry[5,6] and in the production of organic pigments[7-9] have been described.

References

1. 'Cyclohexanone', *Encyclopedia of chemical technology*, 3rd edn, (P. E. Kirk and D. F. Othmer eds), vol 6, p 685. New York: Interscience, 1965.
2. W. Smith and W. G. Reed, 'Some developments in organic chemical manufacture—dyestuffs', *Ind. Chemist*, 1948, **24**(10), 675.
3. G. H. Frank, *The manufacture of intermediates for dyes*. London: Constable, 1950.
4. D. D. W. Adams, 'Intermediates for the dyestuffs and allied industries', *Chemy Ind.*, 1961, 1724.
5. E. E. Abrahart, 'Some recent developments in dyestuffs chemistry', *Chemy Ind.*, 1962, 512.
6. C. V. Stead, 'Recent advances in dyestuffs chemistry', *Chem. Brit.*, 1965, **1**(10), 364.
7. H. Gaertner, 'Modern chemistry of organic pigments', *J. Oil. Col. Chem. Ass*, 1965, **46**(7), 13.
8. N. V. Shah, 'Recent developments in organic pigments', *J. Soc. Dyers Colour.*, 1967, 220.
9. E. R. Innan, 'Organic pigments', *R. Inst. Chem. Lect. Series*, 1967.

Appendix : Prices of Some Common Organic Chemicals

(October 1970; per ton unless otherwise stated)

General chemicals

Acetaldehyde	£122
Acetic acid (glacial— 98/100 per cent)	£73
Acetic anhydride	£103
Calcium carbide	£41
Citric acid (granular)	£220–£237/1000 kg
Citric acid (anhydrous BP)	£230–£247/kg
Copper acetate (30–32 per cent Cu)	£495
Cream of tartar	£11.6/cwt
Formaldehyde	£37.375
Fumaric acid	£217
Gallic acid (BPC)	62.5p/lb
Glycerine (chemically pure)	£11.1/cwt
Iodoform	£1.5/lb
Lactic acid (edible, 80 per cent by weight)	£275
Lead acetate	£330–£370
Maleic anhydride	5p/lb
Malic acid (anhydrous)	£223
Oleic acid	£171
Oxalic acid	£125
Propylene oxide	£148
Salicylic acid	12.5p/lb
Sodium acetate	£54.5
Sodium cyanamide	£6.75/cwt
Sodium lactate (edible, 70 per cent)	5–6p/lb
Sodium prussiate	5–6p/lb
Stearic acid (BPC)	£124–£159
Tartaric acid	£388–£397/1000 kg
Urea	£41.25
Zinc acetate	£282

Coal-tar products

Anthracene (40 per cent)	£2.5/cwt
Benzene (90's)	15–17p/lb
m-Cresol (*meta* content 40–42 per cent)	52.5–58p/gall
Cresol (*meta* content 52–53 per cent)	63–70p/gall
Cresylic acid (99–100 per cent)	47.5p/gall
Naphtha (90–160 °C)	10.5p/gall
Naphtha (90–190 °C)	10.5p/gall

Naphthalene (according to mp)	£22–£30
Naphthalene (hot pressed)	£27.5
Naphthalene (crystals)	£60
Phenol	44p/lb
Pyridine (90–160 °C)	77.5p/gall
Toluene (nitration grade)	10p/gall
Xylene	9.5–10.5p/gall

Dyestuff intermediates

m-Cresol (98–100 per cent)	21.5p/lb
o-Cresol (30–31 °C)	7p/lb
p-Cresol (34–35 °C)	25p/lb
2,5-Dichloroaniline	18p/lb
Dimethylaniline	13p/lb
m-Dinitrobenzene (88–99 °C)	10.5p/lb
Dinitrotoluene (softening point 15 °C)	11p/lb
Dinitrotoluene (softening point 20 °C)	7.5p/lb
Dinitrotoluene (softening point 33 °C)	6p/lb
Dinitrotoluene (softening point 66–68 °C)	10.5p/lb
p-Nitroaniline	18p/lb
Nitrobenzene	4.5p/lb
Nitronaphthalene	12.5p/lb
o-Toluidine	7p/lb
p-Toluidine	14p/lb

Plastics moulding materials

Cellulose acetate	12–15p/lb
Phenol–formaldehyde moulding powder	7.5–20p/lb
Polyethylene (moulding grade)	7p/lb
Polyethylene (extrusion grade)	7.5p/lb
Polyvinyl chloride	8.5–16.5p/lb
Polystyrene	6p/lb

Solvents and plasticizers

Acetic anhydride	£113
Acetone	£63–£83
Adipates (C_7–C_9 alcohols)	£326.50
Adipates (C_8–C_{10} alcohols)	£323
Alcohol (industrial)	20.5–21.5p/gall
Amyl acetate	£251
Amyl lactate	£557
Butyl acetate	£124
n-Butyl alcohol	£125
sec-Butyl alcohol	£82–£178
t-Butyl alcohol	£154.50–£240.50
Butyl lactate	£486
Butyl laurate	£300
Butyl stearate	£250
Carbon disulphide	£61–£67
Carbon tetrachloride	4.5p/lb
Diacetin	£361
Diacetone alcohol	£137–£233

Diallyl phthalate	£436
Dibutyl phthalate	£242.50
Dibutyl glycol phthalate	£375
Dibutyl maleate	£242.50
Di-isobutyl phthalate	£163.50
Dibutyl sebacate	£710
Dibutyl tartrate	£419
Diethyl oxalate	£323
Diethyl phthalate	£185–£193
Dimethyl glycol phthalate	£418
Dimethyl phthalate	£175–£183
Dinonyl maleate	£232.50
Dinonyl phthalate	£155.50
Dinonyl sebacate	£565
Dioctyl adipate	£359
Di-isooctyl maleate	£225
Dioctyl maleate	£284
Di-isooctyl phthalate	£160
Dioctyl phthalate	£194–£205
Di-isooctyl sebacate	£535
Dioctyl sebacate	£575
Dipropylene glycol	£179
Ether	5p/lb
Ethyl acetate	£99
Ethyl alcohol	20–22p/gall
Ethyl alcohol (absolute)	23p/gall
2-Ethylhexanol	£160.50
Ethyl oleate	£310
Hexamine	7.5p/lb
Isobutyl acetate	£105
Isopropyl acetate	£96
Isopropyl alcohol	£88.75–£109.10
Isopropyl oleate	£297
Methanol	£48.50
Methylated spirit (industrial 66 °C)	28–28.5p/gall
Methyl ethyl ketone	£111–£124
Methyl isobutyl carbinol	£133–£235
Methyl isobutyl ketone	£126–£222
Paraformaldehyde	7.5p/lb
Phthalates (C_7–C_9 alcohols)	£160
Polyethylene glycol	£194
Propylene	£182
Styrene	£85
Triacetin	£339
Triamyl citrate	£520
Tributyl citrate	£417
Tricresyl phosphate	£231
Triphenyl phosphate	£341
Trixylyl phosphate	£193

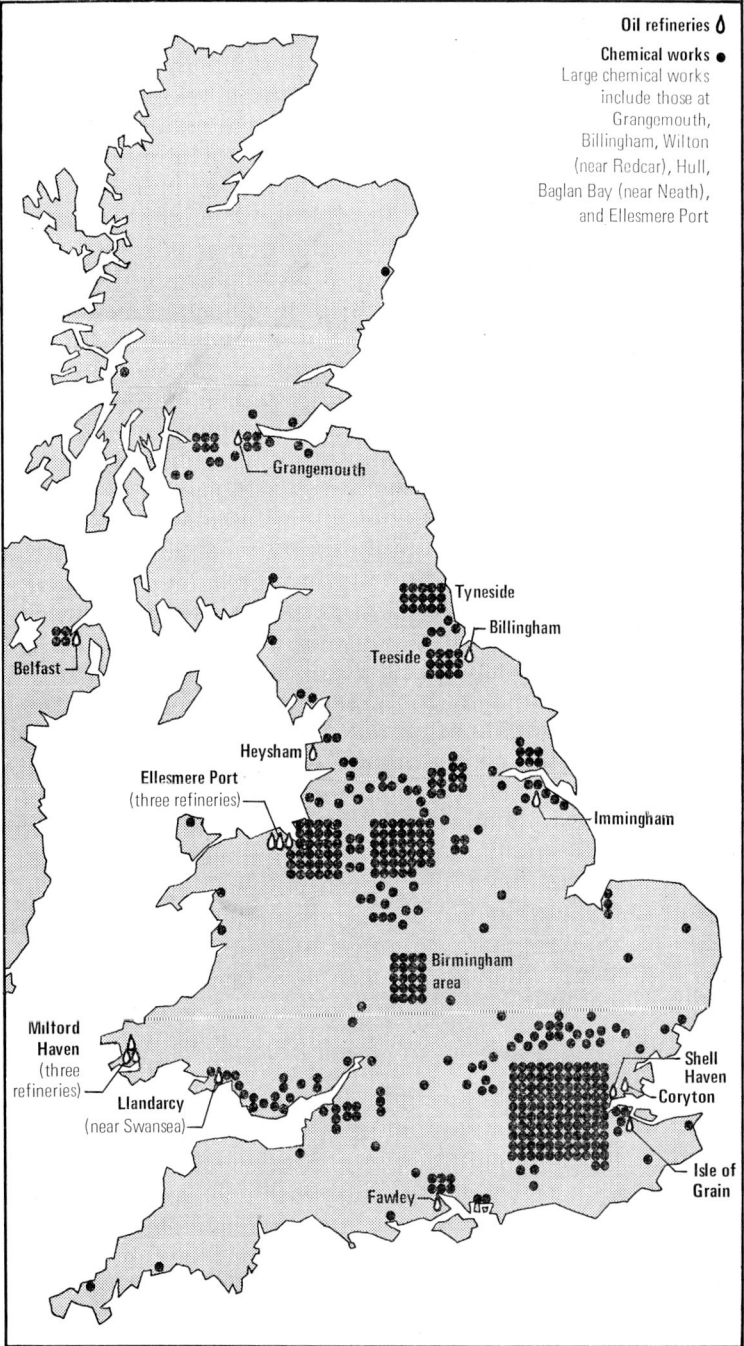

Oil refineries ◊
Chemical works ●
Large chemical works
include those at
Grangemouth,
Billingham, Wilton
(near Redcar), Hull,
Baglan Bay (near Neath),
and Ellesmere Port

Grangemouth

Belfast

Tyneside
Billingham
Teeside

Heysham
Ellesmere Port
(three refineries)
Immingham

Birmingham
area

Milford
Haven
(three
refineries)
Llandarcy
(near Swansea)

Shell
Haven
Coryton

Isle of
Grain

Fawley

Suggestions for Further Reading

Thorpe's Dictionary of applied chemistry, 4th edn. London: Longmans, Green, 1937–1956.

Sources of information on the rubber, plastics and the allied industries. Oxford: Pergamon, 1969.

E. N. Abrahart, *Dyestuffs and intermediates.* Oxford: Pergamon, 1968.

M. J. Astle, *Industrial organic nitrogen compounds.* ACS Monograph No. 150. New York: Reinhold, 1961.

A. Baines, F. R. Bradbury and W. Suckling, *Research in the chemical industry.* London: Elsevier, 1969.

J. A. Brydson, *Plastics materials*, 2nd edn. London: Butterworths, 1969.

W. L. Faith, D. B. Keyes and R. L. Clark, *Industrial chemicals*, 3rd edn. New York: Wiley, 1965.

G. H. Frank, *The manufacture of intermediates and dyes.* London: Constable, 1950.

A. J. Gait, *Heavy organic chemicals.* Oxford: Pergamon, 1967.

R. F. Goldstein and A. L. Waddams, *The petroleum chemicals industry*, 2nd edn. London: E. & E. Spon, 1967.

P. H. Groggins (ed.), *Unit processes in organic synthesis*, 5th edn. New York: McGraw–Hill, 1958.

L. A. Haddock, *Analysis in the chemical industry.* Oxford: Pergamon, 1968.

E. Kilner and D. M. Samuel, *Applied organic chemistry.* London: Macdonald & Evans, 1960.

P. E. Kirk and D. F. Othmer (eds), *Encyclopedia of chemical technology*, 3rd edn. New York: Interscience, 1963.

C. Marsden, *Solvents guide*, 2nd edn. London: Cleaver–Hume, 1963.

D. McNeil, *Coal carbonization products.* Oxford: Pergamon, 1967.

A. Rieche, *Outline of industrial organic chemistry.* London: Butterworths, 1964.

M. Sittig, *Organic chemical process encyclopedia.* Parkridge, N. J.: Noyes, 1967.

W. Mayo Smith (ed.), *Manufacture of plastics*, vols 1 and 2. New York: Reinhold, 1964.

R. M. Stephenson, *Introduction to the chemical process industry.* New York: Reinhold, 1966.

A. L. Waddams, *Chemicals from petroleum*, 2nd edn. London: John Murray, 1968.